THE
MAJOR ACHIEVEMENTS OF
SCIENCE

BY

A. E. E. McKENZIE

VOLUME II

SELECTIONS FROM THE
LITERATURE

CAMBRIDGE
AT THE UNIVERSITY PRESS
1960

PUBLISHED BY
THE SYNDICS OF THE CAMBRIDGE UNIVERSITY PRESS

Bentley House, 200 Euston Road, London, N.W. 1
American Branch: 32 East 57th Street, New York 22, N.Y.

COMMENTARY

©

CAMBRIDGE UNIVERSITY PRESS

1960

Printed in Great Britain at the University Press, Cambridge
(Brooke Crutchley, University Printer)

CONTENTS

v

CONTENTS

CONTENTS

vii

CONTENTS

CONTENTS

ix

PREFACE

This book of extracts, taken mainly from the classics of science, follows the same sequence as my first volume, *The Major Achievements of Science*. The extracts are numbered seriatim, in the order of the chapters in the first volume, and they are arranged under these chapter headings.

They have been selected for the general reader rather than primarily for the science specialist. Their selection rested upon their liveliness and human interest as well as upon their scientific value. Hence only a minority of the extracts have been taken from scientific papers, which are apt to be too technical for the general reader; many have come from books, lectures and letters.

The extracts are usually quite short and, I hope, to the point. But occasionally I have made them rather longer, when the subject is one of general interest: for example the extracts from Lamarck and Darwin on evolution, from Hutton on his geological studies in the Scottish border country, and from Pasteur's dramatic account of his first treatment of victims bitten by rabid dogs.

Some extracts have been selected because they are amusing as well as instructive; these include the account of Archimedes from Plutarch's *Lives*, Swift's attack on astrology, the duologue from Diderot, Voltaire's *conte*, Goethe's conversation with Eckermann about his theory of colours and the article by Raphael Demos. Thomas Young's ponderous reply to the vindictive attacks made upon him in the *Edinburgh Review* can hardly be called an amusing document, but it has its human interest and it includes a notable statement of the principle of interference.

The extracts illustrating the chapters on the seventeenth, eighteenth, nineteenth and twentieth centuries are mainly philosophical. For the seventeenth century, brief extracts from Bacon's *Novum Organum*, Descartes' *Discours de la Méthode* and Hobbes' *Leviathan* are accompanied by a rather longer passage from *New Atlantis*. For the eighteenth century, Locke, Berkeley, Hume and Kant were inevitable choices; something astringent and sparkling from Diderot and Voltaire give the flavour of the Enlightenment; while Goethe represents the school of *Naturphilosophie*. The nineteenth century is represented by Comte, Tyndall, Huxley, Spencer and Haeckel. One could question whether any of them ever produced a classic but they had a wide influence and they illustrate vividly the impact of science on nineteenth-century thought. For twentieth-century philosophy of science, extracts from Mach, Poincaré, Einstein and Eddington were almost obligatory, while the rival philosophy of Russian science is represented by Engels and Lenin.

The sources of the extracts are listed at the end of the book and I wish to express my thanks to the various publishers who have given permission for their reproduction.

A. E. E. M.

CAMBRIDGE
April 1960

CHAPTER I

SCIENCE AND TECHNOLOGY IN ANCIENT AND MEDIEVAL TIMES

1. From ARISTOTLE'S *Physics*

[The following is the opening passage of Aristotle's *Physics* and it illustrates the great achievement of the Greeks in creating science as an abstract, logical system based on principles, replacing the animism of earlier civilizations.]

In all sciences that are concerned with principles or causes or elements, it is acquaintance with these that constitutes knowledge or understanding. For we conceive ourselves to know about a thing when we are acquainted with its ultimate causes and first principles and have got down to its elements. Obviously, then, in the study of Nature too, our first object must be to establish principles.

Now the path of investigation must lie from what is more immediately cognizable and clear to us, to what is clearer and more intimately cognizable in its own nature; for it is not the same thing to be directly accessible to our cognition and to be intrinsically intelligible. Hence, in advancing to that which is intrinsically more luminous and by its nature accessible to deeper knowledge, we must needs start from what is more immediately within our cognition, though in its own nature less fully accessible to understanding.

Now the things most obvious and immediately cognizable by us are concrete and particular, rather than abstract and general; whereas elements and principles are only accessible to us afterwards, as derived from the concrete later when we have analysed them. So we must advance from the concrete whole to the several constituents which it embraces; for it is the concrete whole that is the more readily cognizable by the senses. And by calling the concrete a 'whole' I mean that it embraces in a single complex a diversity of constituent elements, factors, or properties.

2. From PLUTARCH'S *Lives*

[This entertaining account of the machines devised by Archimedes, and their use in the defence of Syracuse, is particularly instructive in its last paragraph, which illustrates one of the reasons why the scientific revolution did not occur in hellenistic times.]

These machines he had designed and contrived, not as matters of any importance, but as mere amusements in geometry; in compliance with King Hiero's desire and request, some little time before, that he should reduce to practice some part of his admirable speculation in science.... Archimedes, however, in writing to King Hiero, whose friend and near relation he was, had stated that given the force, any given weight might be moved, and even boasted, we are told, relying on the strength of demonstration, that if there were another earth, by going into it he could remove this. Hiero being struck with amazement at this, and entreating him to make good this problem by actual experiment, and show some great weight moved by a small engine, he fixed accordingly upon a ship of burden out of the king's arsenal, which could not be drawn out of the dock without great labour and many men; and, loading her with many passengers and a full freight, sitting himself the while far off, with no great endeavour, but only holding the head of the pulley in his hand and drawing the cords by degrees, he drew the ship in a straight line, as smoothly and evenly as if she had been in the sea. The king, astonished at this, and convinced of the power of the art, prevailed upon Archimedes to make him engines accommodated to all the purposes, offensive and defensive, of a siege....

When, therefore, the Romans assaulted the walls in two places at once, fear and consternation stupefied the Syracusans, believing that nothing was able to resist that violence and those forces. But when Archimedes began to ply his engines, he at once shot against the land forces all sorts of missile weapons, and immense masses of stone that came down with incredible noise and violence; against which no man could stand; for they knocked down those upon whom they fell in heaps, breaking all their ranks and files. In the meantime huge poles thrust out from the walls over the ships sunk some by the great weights which they let down from on high upon them; others they lifted up into the air by an iron hand or beak like a crane's beak, and, when they had drawn them up by the prow, and set them on end upon the poop, they plunged them to the bottom of the sea; or else the ships, drawn by engines within, and whirled about, were dashed against steep rocks that stood jutting out under the walls, with great destruction of the soldiers that were aboard them. A ship was frequently lifted up to a great height in the air (a dreadful thing to behold), and was rolled to and fro, and kept swinging, until the mariners were all thrown out, when at length it was dashed against the rocks, or let fall....

In fine, when such terror had seized upon the Romans, that, if they did but see a little rope or a piece of wood from the wall, instantly crying out, that there it was again, Archimedes was about to let fly some engine at them, they turned their backs and fled, Marcellus desisted from conflicts and assaults, putting all his hope in a long siege. Yet Archimedes possessed so high a spirit, so profound a soul, and such treasures of scientific knowledge, that though these inventions had now obtained him the renown of more than human sagacity, he yet would not deign to leave behind him any commentary or writing on such subjects; but, repudiating as sordid and ignoble the whole trade of engineering, and every sort of art that lends itself to mere use and profit, he placed his whole affection and ambition in those purer speculations where there can be no reference to the vulgar needs of life.

3. From *The Book of Beasts* (a twelfth-century bestiary)

[This extract illustrates the medieval view that everything in nature is symbolical of the spiritual life. Bestiaries were based on the compilation of an anonymous person called Physiologus, who lived between the second and fifth centuries A.D., probably in Egypt.]

Scientists say that Leo [the Lion] has three principal characteristics.

His first feature is that he loves to saunter on the tops of mountains. Then, if he should happen to be pursued by hunting men, the smell of the hunters reaches up to him, and he disguises his spoor behind him with his tail. Thus the sportsmen cannot track him.

It was in this way that our Saviour (i.e. the Spiritual Lion of the Tribe of Judah, the Rod of Jesse, the Lord of Lords, the Son of God) once hid the spoor of his love in the high places, until being sent by the Father, he came down into the womb of the Virgin Mary and saved the human race which had perished. Ignorant of the fact that his spoor could be concealed, the Devil (i.e. the hunter of humankind) dared to pursue him with temptations like a mere man. Even the angels themselves who were on high, not recognizing his spoor, said to those who were going up with him when he ascended to his reward: 'Who is this King of Glory?'

The Lion's second feature is, that when he sleeps, he seems to keep his eyes open.

In this very way, Our Lord also while sleeping in the body, was buried after being crucified—yet his Godhead was awake. As it is said in the *Song of Songs*, 'I am asleep and my heart is awake', or, in the Psalm, 'Behold, he that keepeth Israel shall neither slumber nor sleep'.

The third feature is this, that when a lioness gives birth to her cubs, she brings them forth dead and lays them up lifeless for three days—until their father, coming on the third day, breathes in their faces and makes them alive.

Just so did the Father Omnipotent raise Our Lord Jesus Christ from the dead on the third day. Quoth Jacob: 'He shall sleep like a lion, and the lion's whelp shall be raised.'

4. From JONATHAN SWIFT'S *Predictions for the Year 1708 by Isaac Bickerstaff Esq.*

[After thousands of years of credulity about astrology educated opinion ceased to take it seriously as a result of the new scientific outlook on the universe of the seventeenth century.]

My first prediction is but a trifle, yet I will mention it to show how ignorant these sottish pretenders to astrology are in their own concerns; it relates to Partridge the Almanackmaker. I have consulted the star of his nativity, by my own rules, and find that he will infallibly die upon 29th March next, about eleven at night, of a raging fever; therefore advise him to consider of it and settle his affairs in time.

[When Mr Partridge protested that he had not died, Swift retorted that over a thousand gentlemen had looked at Partridge's almanack and exclaimed, NO MAN ALIVE EVER WRIT SUCH DAMNED STUFF AS THIS.]

CHAPTER 2

THE COPERNICAN THEORY

5. From Nicolaus Copernicus' *On the Revolutions of the Celestial Orbs* (1543)

[Book I of *On the Revolutions* gives a brief general account of the heliocentric universe. In the following extract Copernicus first considers the diurnal rotation of the Earth and suggests that the Earth would not fly to pieces under the action of the centrifugal force caused by its rotation because its motion is natural and not forced. He then explains how the Earth, with the Moon, revolves round the Sun, together with the rest of the Planets, whose distances from the Sun must be in the order of their periods of revolution. Saturn, with a period of 30 years, is furthest away from the Sun, and Mercury, with a period of 80 days, is nearest. The Sun, in the centre, is described as the Ruler of the Universe.]

If then, says Ptolemy, Earth moves at least with a diurnal rotation, the result must be the reverse of that described above [i.e. that terrestial elements move in straight lines up or down]. For the motion must be of excessive rapidity, since in 24 hours it must impart a complete rotation of the Earth. Now things rotating very rapidly resist cohesion or, if united, are apt to disperse, unless firmly held together. Ptolemy therefore says that Earth would have been dissipated long ago, and (which is the height of absurdity) would have destroyed the Heavens themselves; and certainly all living creatures and other heavy bodies free to move could not have remained on its surface, but must have been shaken off. Neither could falling objects reach their appointed place vertically beneath, since in the meantime the Earth would have moved swiftly from under them. Moreover clouds and everything in the air would continually move westward.

The Insufficiency of these Arguments, and their Refutation

For these and like reasons, they say that Earth surely rests at the centre of the Universe. Now if one should say that the Earth *moves*, that is as much as to say that the motion is natural, not

5

forced; and things which happen according to nature produce the opposite effects to those due to force. Things subject to any force, gradual or sudden, must be disintegrated, and cannot long exist. But natural processes being adapted to their purpose work smoothly.

Idle therefore is the fear of Ptolemy that Earth and all thereon would be disintegrated by a natural rotation, a thing far different from an artificial act....

Why then hesitate to grant Earth that power of motion natural to its shape, rather than suppose a gliding round of the whole Universe, whose limits are unknown and unknowable? And why not grant that the diurnal rotation is only apparent in the Heavens but real in the Earth? It is but as the saying of Aeneas in Virgil— 'We sail forth from the harbour, and lands and cities retire'. As the ship floats along in the calm, all external things seem to have the motion that is really that of the ship, while those within the ship feel that they and all its contents are at rest.

*   *   *

Since then there is no reason why the Earth should not possess the power of motion, we must consider whether in fact it has more motions than one, so as to be reckoned as a Planet.

That the Earth is not the centre of all revolutions is proved by the apparently irregular motions of the Planets and the variations in their distances from the Earth. These would be unintelligible if they moved in circles concentric with the Earth. Since, therefore, there are more centres than one, we may discuss whether the centre of the Universe is or is not the Earth's centre of gravity.

Now it seems to me gravity is but a natural inclination, bestowed on the parts of bodies by the Creator so as to combine the parts in the form of a sphere and thus contribute to their unity and integrity. And we may believe this property present even in the Sun, Moon and Planets, so that thereby they retain their spherical form notwithstanding their various paths. If, therefore, the Earth also has other motions, these must necessarily resemble the many outside motions having a yearly period. For if we transfer the motion of the Sun to the Earth, taking the Sun to be at rest, then morning and evening, risings and settings of Stars will be unaffected, while the stationary points, retrogressions, and progressions of the Planets are due not to their own proper motions, but to that of the Earth, which they reflect. Finally, we shall place the Sun himself at the centre of the Universe. All this is suggested by the systematic procession of events and the harmony of the whole Universe, if only we face the facts, as they say, 'with both eyes open'. . . .

We therefore assert that the centre of the Earth, carrying the Moon's path, passes in a great orbit among the other Planets in an annual revolution round the Sun; that near the Sun is the centre of the Universe; and that whereas the Sun is at rest, any apparent motion of the Sun can be better explained by the motion of the Earth. Yet so great is the Universe that though the distance of the Earth from the Sun is not insignificant compared with the size of any other planetary path, in accordance with the ratios of their sizes, it is insignificant compared with the distance of the sphere of the Fixed Stars.

I think it is easier to believe this than to confuse the issue by assuming a vast number of spheres, which those who keep Earth at the centre must do. We thus rather follow Nature, who producing nothing vain or superfluous often prefers to endow one cause with many effects. Though these views are difficult, contrary to expectation, yet in the sequel we shall, God willing, make them abundantly clear at least to mathematicians.

Given the above view—and there is none more reasonable—that the periodic times are proportional to the sizes of the orbits, then the order of the spheres, beginning with the most distant, is as follows. Most distant of all is the Sphere of the Fixed Stars, containing all things, and being therefore itself immovable. It represents that to which the motion and position of all the other bodies must be referred. Some hold that it too changes in some way, but we shall assign another reason for this apparent change, as will appear in the account of the Earth's motion. Next is the planet Saturn, revolving in 30 years. Next comes Jupiter, moving in a 12 year circuit: then Mars, who goes round in 2 years. The fourth place is held by the annual revolution in which the Earth is contained, together with the orbit of the Moon as on an epicycle. Venus, whose period is 9 months, is in the fifth place, and sixth is Mercury, who goes round in the space of 80 days.

In the middle of all sits Sun enthroned. In this most beautiful temple could we place this luminary in any better position from

7

which he can illuminate the whole at once? He is rightly called the Lamp, the Mind, the Ruler of the Universe; Hermes Trismegistus names him the Visible God, Sophocles' Electra calls him the All-Seeing. So the Sun sits as upon a royal throne ruling his children the planets which circle round him.

6. From GALILEO'S letter to Madame CHRISTINA OF LORRAINE, Grand Duchess of Tuscany, *Concerning the Use of Biblical Quotations in Matters of Science* (1615)

Some years ago, as Your Serene Highness well knows, I discovered in the heavens many things that had not been seen before our own age. The novelty of these things, as well as some consequences which followed from them in contradiction to the physical notions commonly held among academic philosophers, stirred up against me no small number of professors—as if I had placed these things in the sky with my own hands in order to upset nature and overturn the sciences. . . .

Persisting in their original resolve to destroy me and everything mine by any means they can think of, these men are aware of my views in astronomy and philosophy. They know that as to the arrangement of the parts of the universe, I hold the sun to be situated motionless in the centre of the revolution of the celestial orbs while the earth rotates on its axis and revolves about the sun. They know also that I support this position not only by refuting the arguments of Ptolemy and Aristotle, but by producing many counter-arguments; in particular, some which relate to physical effects whose causes can perhaps be assigned in no other way. In addition there are astronomical arguments derived from many things in my new celestial discoveries that plainly confute the Ptolemaic system while admirably agreeing with and confirming the contrary hypothesis. Possibly because they are disturbed by the known truth of other propositions of mine which differ from those commonly held, and therefore mistrusting their defence so long as they confine themselves to the field of philosophy, these men have resolved to fabricate a shield for their fallacies out of the mantle of pretended religion and the authority of the Bible. These they apply, with little judgment, to the refutation of arguments that they do not understand and have not even listened to.

First they have endeavoured to spread the opinion that such propositions in general are contrary to the Bible and are consequently damnable and heretical....

With regard to this argument, I think in the first place that it is very pious to say and prudent to affirm that the holy Bible can never speak untruth—whenever its true meaning is understood. But I believe nobody will deny that it is often very abstruse, and may say things which are quite different from what its bare words signify. Hence in expounding the Bible if one were always to confine oneself to the unadorned grammatical meaning, one might fall into error. Not only contradictions and propositions far from true might thus be made to appear in the Bible, but even grave heresies and follies. Thus it would be necessary to assign to God feet, hands, and eyes, as well as corporeal and human affections, such as anger, repentance, hatred, and sometimes even the forgetting of things past and ignorance of those to come. These propositions uttered by the Holy Ghost were set down in that manner by the sacred scribes in order to accommodate them to the capacities of the common people, who are rude and unlearned. For the sake of those who deserve to be separated from the herd, it is necessary that wise expositors should produce the true senses of such passages, together with the special reasons for which they were set down in these words. This doctrine is so widespread and so definite with all theologians that it would be superfluous to adduce evidence for it.

7. From GALILEO'S *Dialogue concerning the Two Chief World Systems—Ptolemaic and Copernican* (1632)

[There are three interlocutors: Salviati, who represents the views of Galileo; Simplicio, who is a follower of Aristotle; and Sagredo who represents an intelligent layman, uncommitted to either side. The extract is taken from The Third Day, about three-quarters of the way through the Dialogue, and in it the arrangement of the planets round the Sun is discussed in some detail.]

Salviati. Now if it is true that the centre of the universe is that point around which all the orbs and world bodies (that is, the planets) move, it is quite certain that not the earth, but the sun, is to be found at the centre of the universe. Hence, as for this first general conception, the central place is the sun's, and the earth is to be found as far away from the centre as it is from the sun.

Simplicio. How do you deduce that it is not the earth, but the sun, which is at the centre of the revolutions of the planets?

Salviati. This is deduced from most obvious and therefore most powerfully convincing observations. The most palpable of these, which excludes the earth from the centre and places the sun there, is that we find all the planets closer to the earth at one time and further from it at another. The differences are so great that Venus, for example, is six times as distant from us at its farthest as at its closest, and Mars soars nearly eight times as high in the one state as in the other. You may thus see whether Aristotle was not some trifle deceived in believing that they were always equally distant from us.

Simplicio. But what are the signs that they move around the sun?

Salviati. This is reasoned out from finding the three outer planets—Mars, Jupiter, and Saturn—always quite close to the earth when they are in opposition to the sun, and very distant when they are in conjunction with it.[1] This approach and recession is of such moment that Mars when close looks sixty times as large as when it is most distant. Next, it is certain that Venus and Mercury must revolve around the sun, because of their never moving far away from it, and because of their being seen now beyond it and now on this side of it, as Venus's changes of shape conclusively prove. As to the moon, it is true that this can never separate from the earth in any way, for reasons that will be set forth more specifically as we proceed.

Sagredo. I have hopes of hearing still more remarkable things arising from this annual motion of the earth than were those which depended upon its diurnal rotation.

Salviati. You will not be disappointed, for as to the action of the diurnal motion upon celestial bodies, it was not and could not be anything different from what would appear if the universe were to rush speedily in the opposite direction. But this annual motion, mixing with the individual motions of all the planets, produces a great many oddities, which in the past have baffled all the greatest men in the world.

[1] Heavenly bodies in opposition differ in position by 180°, and in conjunction they have the same longitude, i.e. are passing each other.

Now returning to these first general conceptions, I repeat that the centre of the celestial rotation for the five planets, Saturn, Jupiter, Mars, Venus, and Mercury, is the sun; this will hold for the earth too, if we are successful in placing that in the heavens. Then as to the moon, it has a circular motion around the earth, from which as I have already said it cannot be separated; but this does not keep it from going around the sun along with the earth in its annual movement.

Simplicio. I am not yet convinced of this arrangement at all. Perhaps I should understand it better from the drawing of a diagram, which might make it easier to discuss.

Salviati. That shall be done. But for your greater satisfaction and your astonishment, too, I want you to draw it yourself. You will see that however firmly you may believe yourself not to understand it, you do so perfectly, and just by answering my questions you will describe it exactly. So take a sheet of paper and the compasses; let this page be the enormous expanse of the universe, in which you have to distribute and arrange its parts as reason shall direct you. And first, since you are sure without my telling you that the earth is located in this universe, mark some point at your pleasure where you intend this to be located, and designate it by means of some letter.

Simplicio. Let this be the place of the terrestrial globe, marked *A*.

Salviati. Very well. I know in the second place that you are aware that this earth is not inside the body of the sun, nor even contiguous to it, but is distant from it by a certain space. Therefore assign to the sun some other place of your choosing, as far from the earth as you like, and designate that also.

Simplicio. Here I have done it; let this be the sun's position, marked *O*.

Salviati. These two established, I want you to think about placing Venus in such a way that its position and movement can conform to what sensible experience shows us about it. Hence you must call to mind, either from past discussions or from your own observations, what you know happens with this star. Then assign it whatever place seems suitable for it to you.

Simplicio. I shall assume that those appearances are correct which you have related and which I have read also in the booklet of theses; that is, that this star never recedes from the sun beyond a certain definite interval of forty degrees or so; hence it not only never reaches opposition to the sun, but not even quadrature, nor so much as a sextile aspect.[1] Moreover, I shall assume that it displays itself to us about forty times as large at one time than at another; greater when, being retrograde, it is approaching evening conjunction with the sun, and very small when it is moving forward toward morning conjunction, and furthermore that when it appears very large, it reveals itself in a horned shape, and when it looks very small it appears perfectly round.

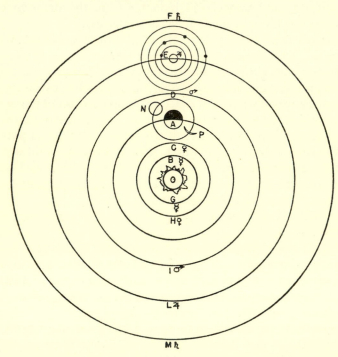

These appearances being correct, I say, I do not see how to escape affirming that this star revolves in a circle around the sun, in such a way that this circle cannot possibly be said to embrace and contain within itself the earth, nor to be beneath the sun (that is, between the sun and the earth), nor yet beyond the sun. Such a circle cannot embrace the earth because then Venus would

[1] Quadrature means that heavenly bodies differ in position by 90°, sextile aspect by 60°.

sometimes be in opposition to the sun; it cannot be beneath the sun, for then Venus would appear sickle-shaped at both conjunctions; and it cannot be beyond the sun, since then it would always look round and never horned. Therefore for its lodging I shall draw the circle *CH* around the sun, without having this include the earth.

Salviati. Venus provided for, it is fitting to consider Mercury, which, as you know, keeping itself always around the sun, recedes therefrom much less than Venus. Therefore consider what place you should assign to it.

Simplicio. There is no doubt that, imitating Venus as it does, the most appropriate place for it will be a smaller circle, within this one of Venus and also described about the sun. A reason for this, and especially for its proximity to the sun, is the vividness of Mercury's splendour surpassing that of Venus and all the other planets. Hence on this basis we may draw its circle here and mark it with the letters *BG*.

Salviati. Next, where shall we put Mars?

Simplicio. Mars, since it does come into opposition with the sun, must embrace the earth with its circle. And I see that it must also embrace the sun; for, coming into conjunction with the sun, if it did not pass beyond it but fell short of it, it would appear horned as Venus and the moon do. But it always looks round; therefore its circle must include the sun as well as the earth. And since I remember your having said that when it is in opposition to the sun it looks sixty times as large as when in conjunction, it seems to me that this phenomenon will be well provided for by a circle around the sun embracing the earth, which I draw here and mark *DI*. When Mars is at the point *D*, it is very near the earth and in opposition to the sun, but when it is at the point *I*, it is in conjunction with the sun and very distant from the earth.

And since the same appearances are observed with regard to Jupiter and Saturn (although with less variation in Jupiter than in Mars, and with still less in Saturn than in Jupiter), it seems clear to me that we can also accommodate these two planets very neatly with two circles, still around the sun. This first one, for Jupiter, I mark *EL*; the other, higher, for Saturn, is called *FM*.

Salviati. So far you have comported yourself uncommonly well. And since, as you see, the approach and recession of the three outer planets is measured by double the distance between the earth and the sun, this makes a greater variation in Mars than in Jupiter because the circle *DI* of Mars is smaller than the circle *EL* of Jupiter. Similarly, *EL* here is smaller than the circle *FM* of Saturn, so the variation is still less in Saturn than in Jupiter, and this corresponds exactly to the appearances. It now remains for you to think about a place for the moon.

Simplicio. Following the same method (which seems to me very convincing), since we see the moon come into conjunction and opposition with the sun, it must be admitted that its circle embraces the earth. But it must not embrace the sun also, or else when it was in conjunction it would not look horned but always round and full of light. Besides, it would never cause an eclipse of the sun for us, as it frequently does, by getting in between us and the sun. Thus one must assign to it a circle around the earth, which shall be this one, *NP*, in such a way that when at *P* it appears to us here on the earth *A* as in conjunction with the sun, which sometimes it will eclipse in this position. Placed at *N*, it is seen in opposition to the sun, and in that position it may fall under the earth's shadow and be eclipsed.

Salviati. Now what shall we do, Simplicio, with the fixed stars? Do we want to sprinkle them through the immense abyss of the universe, at various distances from any predetermined point, or place them on a spherical surface extending around a centre of their own so that each of them will be the same distance from that centre?

Simplicio. I had rather take a middle course, and assign to them an orb described around a definite centre and included between two spherical surfaces—a very distant concave one, and another closer and convex, between which are placed at various altitudes the innumerable host of stars. This might be called the universal sphere, containing within it the spheres of the planets which we have already designated.

Salviati. Well, Simplicio, what we have been doing all this while is arranging the world bodies according to the Copernican distribution, and this has now been done by your own hand.

8. From letters written by JOHANNES KEPLER to HERWART

[In the first extract Kepler states his intention to find a mechanical interpretation of the universe. In the second he explains that he favours the
Copernican theory for mechanical or physical reasons.]

Prague, 10 February 1605

My aim is to show that the heavenly machine is not a kind of
divine, live being, but a kind of clockwork (and he who believes
that a clock has a soul, attributes the maker's glory to the work),
insofar as nearly all the manifold motions are caused by a most
simple, magnetic, and material force, just as all motions of the
clock are caused by a simple weight. And I also show how these
physical causes are to be given numerical and geometrical
expression.

Prague, 28 March 1605

You ask me, Magnificence, about the hypotheses of Copernicus
and you seem to be pleased that I insist on my opinion....

[One of my main ideas aimed against Tycho is] if the sun moves
round the earth, then it must, of necessity, along with the other
planets become sometimes faster, sometimes slower in its movements, and this without following fixed courses, since there are
none. But this is incredible. Furthermore, the sun which is so much
higher ranking than the unimportant earth would have to be moved
by the earth in the same way as the five other planets are put in
motion by the sun. That is completely absurd. Therefore it is
much more plausible that the earth together with the five planets
is put in motion by the sun and only the moon by the earth.

CHAPTER 3

THE MECHANICAL UNIVERSE

9. From ARISTOTLE'S *Physics*

[This extract explains why light bodies rise and heavy bodies fall, in their motion to their 'proper place'. It illustrates Aristotle's thesis that all movement is change from potential to actual existence, and that Nature is essentially teleological, that is, it has inherent properties tending to a goal or purpose. Galileo and Newton completely rejected this thesis in their creation of modern physics.]

If the question is still pressed why light and heavy things tend to their respective positions, the only answer is that they are natured so, and that what we mean by heavy and light as distinguished and defined is just this downward or upward tendency. As we have said, here too there are different stages of potentiality. When a substance is water it is already in a way potentially light, and when it is air it may still be only potentially in the position proper to it, for its ascent may be hindered, but if the hindrance be removed it actualizes the potentiality and continuously mounts. And likewise the potentially 'such' tends to its actual realization; even as knowledge becomes straightway active if not impeded; and so likewise are the potential dimensions of a thing realized if there be no hindrance. If anyone removes the obstacle he may be said in one sense (but in another not) to cause the movement; for instance if he removes a column from beneath the weight it was supporting, or cuts the string that attached a bladder, under water, to the stone that holds it down, for he incidentally determines the moment at which the potential motion becomes actual, just as the wall from which a ball rebounds determines the direction in which the ball rebounds, though it is the player and not the wall that is the cause of its motion. So it is now clear that in no one of these cases does the thing that is in motion move itself; but it has the passive (though not the active and efficient) principle of movement inherent in itself.

10. From GALILEO'S *Dialogue concerning the Two Chief World Systems—Ptolemaic and Copernican* (1632)

[This is an example of the kind of argument by which Galileo endeavoured to refute the doctrine of Aristotle given in the preceding extract. Salviati gives Galileo's views and Simplicio is an Aristotelian.]

Salviati. I consider the upward motion of heavy bodies due to received impetus to be just as natural as their downward motion dependent upon gravity.

Simplicio. This I shall never admit, because the latter has a natural and perpetual internal principle while the former has a finite and constrained external one.

Salviati. If you flinch from conceding to me that the principles of motion of heavy bodies downward and upward, are equally internal and natural, what would you do if I were to tell you that they may be also one and the same?

Simplicio. I leave it to you to judge.

Salviati. Rather, I want you to be the judge. Tell me, do you believe that contradictory internal principles can reside in the same natural body?

Simplicio. Absolutely not.

Salviati. What would you consider to be the natural intrinsic tendencies of earth, lead and gold, and in brief of all very heavy materials? That is, towards what motion do you believe that their internal principle draws them?

Simplicio. Motion towards the centre of heavy things; that is, to the centre of the universe and of the earth, whither they would be conducted if not impeded.

Salviati. So that if the terrestrial globe were pierced by a hole which passed through its centre, a cannon ball dropped through this and moved by its natural and intrinsic principle would be taken to the centre, and all this motion would be spontaneously made and by an intrinsic principle. Is that right?

Simplicio. I take that to be certain.

Salviati. But having arrived at the centre is it your belief that it would pass on beyond, or that it would immediately stop its motion then?

Simplicio. I think it would keep on going a long way.

Salviati. Now wouldn't this motion beyond the centre be upward, and according to what you have said preternatural and constrained? But upon what other principle will you make it depend, other than the very one which has brought the ball to the centre and which you have already called intrinsic and natural?...

Simplicio. I believe there are answers to all these objections, though for the moment I do not remember them.

11. From GALILEO'S *Dialogues concerning Two New Sciences* (1638)

FOURTH DAY

Salviati. Once more, Simplicio is here on time; so let us without delay take up the question of motion. The text of our Author [Galileo] is as follows:

The Motion of Projectiles

In the preceding pages we have discussed the properties of uniform motion and of motion naturally accelerated along planes of all inclinations. I now propose to set forth those properties which belong to a body whose motion is compounded of two other motions, namely, one uniform and one naturally accelerated; these properties, well worth knowing, I propose to demonstrate in a rigid manner. This is the kind of motion seen in a moving projectile; its origin I conceive to be as follows:

Imagine any particle projected along a horizontal plane without friction; then we know, from what has been more fully explained in the preceding pages, that this particle will move along this same plane with a motion which is uniform and perpetual, provided the plane has no limits. But if the plane is limited and elevated, then the moving particle, which we imagine to be a heavy one, will on passing over the edge of the plane acquire, in addition to its previous uniform and perpetual motion, a downward propensity due to its own weight; so that the resulting motion which I call projection (*projectio*), is compounded of one which is uniform and horizontal and of another which is vertical and naturally accelerated. We now proceed to demonstrate some of its properties, the first of which is as follows:

Theorem I, Proposition I

A projectile which is carried by a uniform horizontal motion compounded with a naturally accelerated vertical motion describes a path which is a semi-parabola.

Sagredo. Here, Salviati, it will be necessary to stop a little while for my sake and, I believe, also for the benefit of Simplicio; for it so happens that I have not gone very far in my study of Apollonius and am merely aware of the fact that he treats of the parabola and other conic sections, without an understanding of which I hardly think one will be able to follow the proof of other propositions depending upon them. Since even in this first beautiful theorem the author finds it necessary to prove that the path of a projectile is a parabola, and since, as I imagine, we shall have to deal with only this kind of curves, it will be absolutely necessary to have a thorough acquaintance, if not with all the properties which Apollonius has demonstrated for these figures, at least with those which are needed for the present treatment.

12. From ISAAC NEWTON'S *Principia* (1687)

[The famous Laws of Motion in this extract come near the beginning of the *Principia*, immediately after the Definitions, indicating the similarity in plan of the work to that of Euclid. Much has been written about the true meaning and implications of the laws, on which classical physics is built.]

AXIOMS, OR LAWS OF MOTION

Law I

Every body continues in its state of rest, or of uniform motion in a right line, unless it is compelled to change that state by forces impressed upon it.

Projectiles continue in their motions, so far as they are not retarded by the resistance of the air, or impelled downwards by the force of gravity. A top, whose parts by their cohesion are continually drawn aside from rectilinear motions, does not cease its rotation, otherwise than as it is retarded by the air. The greater bodies of the planets and comets, meeting with less resistance in freer spaces, preserve their motions both progressive and circular for a much longer time.

Law II

The change of motion is proportional to the motive force impressed; and is made in the direction of the right line in which that force is impressed.

If any force generates a motion, a double force will generate double the motion, a triple force triple the motion, whether that force be impressed altogether and at once, or gradually and successively. And this motion (being always directed the same way with the generating force), if the body moved before, is added to or subtracted from the former motion, according as they directly conspire with or are directly contrary to each other; or obliquely joined, when they are oblique, so as to produce a new motion compounded from the determination of both.

Law III

To every action there is always opposed an equal reaction; or, the mutual actions of two bodies upon each other are always equal and directed to contrary parts.

Whatever draws or presses another is as much drawn or pressed by that other. If you press a stone with your finger, the finger is also pressed by the stone. If a horse draws a stone tied to a rope, the horse (if I may so say) will be equally drawn back towards the stone; for the distended rope, by the same endeavour to relax or unbend itself, will draw the horse as much towards the stone as it does the stone towards the horse, and will obstruct the progress of the one as much as it advances that of the other. If a body impinge upon another, and by its force change the motion of the other, that body also (because of the equality of the mutual pressure) will undergo an equal change, in its own motion, towards the contrary part. The changes made by these actions are equal, not in the velocities but in the motions of bodies; that is to say, if the bodies are not hindered by any other impediments. For, because the motions are equally changed, the changes of the velocities made towards contrary parts are inversely proportional to the bodies. This law takes place also in attractions, as will be proved in the next Scholium.

[Newton here explains that the cause of the revolution of a planet is the same as that of the fall of a stone. The centripetal force to which he refers is gravitational attraction.]

The principle of circular motion in free spaces

After this time, we do not know in what manner the ancients explained the question, how the planets came to be retained within certain bounds in these free spaces, and to be drawn off from the rectilinear courses, which, left to themselves, they should have pursued, into regular revolutions in curvilinear orbits. Probably it was to give some sort of satisfaction to this difficulty that solid orbs had been introduced.

The later philosophers pretend to account for it either by the action of certain vortices, as *Kepler* and *Descartes*; or by some other principle of impulse or attraction as *Borelli, Hooke,* and others of our nation; for, from the laws of motion, it is most certain that these effects must proceed from the action of some force or other.

But our purpose is only to trace out the quantity and properties of this force from the phaenomena, and to apply what we discover in some simple cases as principles, by which, in a mathematical way, we may estimate the effects thereof in more involved cases; for it would be endless and impossible to bring every particular to direct and immediate observation.

We said, in a mathematical way, to avoid all questions about the nature or quality of this force, which we would not be understood to determine by any hypothesis; and therefore call it by the general name of a centripetal force, as it is a force which is directed towards some centre; and as it regards more particularly a body in that centre, we call it circumsolar, circumterrestrial, circumjovial; and so in respect of other central bodies.

The action of centripetal forces

That by means of centripetal forces the planets may be retained in certain orbits, we may easily understand, if we consider the motions of projectiles; for a stone that is projected is by the pressure of its own weight forced out of the rectilinear path, which by the initial projection alone it should have pursued, and made to describe a curved line in the air; and through that crooked way is at last brought down to the ground; and the greater the velocity is with which it is projected, the farther it goes before it falls to the earth. We may therefore suppose the velocity to be so increased that it would describe an arc of 1, 2, 5, 10, 100, 1000 miles before it arrived at the earth, till at last, exceeding the limits of the earth, it should pass into space without touching it.

Let *AFB* represent the surface of the earth, *C* its centre, *VD, VE, VF* the curved lines which a body would describe, if projected in an horizontal direction from the top of an high mountain successively with more and more velocity; and, because the celestial motions are scarcely retarded by the little or no resistance of the spaces in which they are performed, to keep up the parity of cases, let us suppose either that there is no air about the earth, or at least that it is endowed with little or no power of resisting; and for the same reason that the body projected with a less velocity describes the lesser arc *VD*, and with a greater velocity the greater arc *VE*, and augmenting the velocity, it goes farther

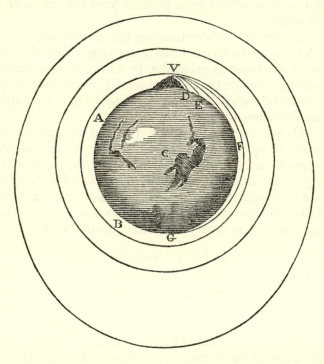

and farther to *F* and *G*, if the velocity was still more and more augmented, it would reach at last quite beyond the circumference of the earth, and return to the mountain from which it was projected.

And since the areas which by this motion it describes by a radius drawn to the centre of the earth are (by Prop. 1 Book 1, Princip. Math.) proportional to the times in which they are described, its velocity, when it returns to the mountain will be no less than it was at first; and, retaining the same velocity, it will describe the same curve over and over, by the same law.

But if we now imagine bodies to be projected in the directions of lines parallel to the horizon from greater heights, as of 5, 10, 100, 1000, or more miles, or rather as many semidiameters of the earth, those bodies, according to their different velocity, and the different force of gravity in different heights, will describe arcs either concentric with the earth, or variously eccentric, and go on revolving through the heavens in those orbits just as the planets do in their orbits.

[This well-known passage comes from the General Scholium at the end of Book III. In it Newton states the objective of science. Following the lead given by Galileo, he maintains that science must be experimental and inductive, in contrast to the deductive method of Aristotle. He says that he frames no hypotheses. The hypotheses he is condemning in particular, are those of Aristotle's occult qualities and of Descartes' mechanical vortices. Nevertheless hypothesis has an important place in scientific method and Newton himself showed an extraordinary imaginative fertility in his Queries at the end of his *Opticks*.]

But hitherto I have not been able to discover the cause of those properties of gravity from phaenomena, and I frame no hypotheses; for whatever is not deduced from the phaenomena is to be called an hypothesis; and hypotheses, whether metaphysical or physical, whether of occult qualities or mechanical, have no place in experimental philosophy. In this philosophy particular propositions are inferred from the phaenomena, and afterwards rendered general by induction. Thus it was that the impenetrability, the mobility, and the impulsive force of bodies, and the laws of motion and of gravitation, were discovered. And to us it is enough that gravity does really exist, and act according to the laws which we have explained, and abundantly serves to account for all the motions of celestial bodies, and of our sea.

CHAPTER 4

THE CIRCULATION OF THE BLOOD

13. From ROBERT BOYLE'S *A Disquisition about the Final Causes of Natural Things* (1688)

And I remember, that when I asked our famous Harvey, in the only discourse I had with him, (which was but a while before he died) what were the things, that induced him to think of a circulation of the blood? he answered me, that when he took notice, that the valves in the veins of so many parts of the body were so placed, that they gave free passage of the blood towards the heart, but opposed the passage of the venal blood the contrary way; he was invited to imagine. that so provident a cause as nature had not so placed so many valves without design; and no design seemed more probable, than that since the blood could not well, because of the interposing valves, be sent by the veins to the limbs, it should be sent through the arteries, and return through the veins, whose valves did not oppose its course that way.

14. From WILLIAM HARVEY'S *An Anatomical Disquisition on the Motion of the Heart and Blood in Animals* (1628)

[This extract is taken from the central chapter in Harvey's book. In it he states the essence of his theory, that the blood, impelled by the heart, moves 'as it were, in a circle'. He finds analogies in the motions of the heavenly bodies and in the cycle of evaporation and condensation of the earth's moisture.]

Thus far I have spoken of the passage of the blood from the veins into the arteries, and of the manner in which it is transmitted and distributed by the action of the heart; points to which some, moved either by the authority of Galen or Columbus, or the reasonings of others, will give in their adhesion. But what remains to be said upon the quantity and source of the blood

24

which thus passes, is of so novel and unheard-of character, that I not only fear injury to myself from the envy of a few, but I tremble lest I have mankind at large for my enemies, so much doth wont and custom, that become as another nature, and doctrine once sown and that hath struck deep root, and respect for antiquity influence all men: Still the die is cast, and my trust is in my love of truth, and the candour that inheres in cultivated minds. And sooth to say, when I surveyed my mass of evidence, whether derived from vivisections, and my various reflections on them, or from the ventricles of the heart and the vessels that enter into and issue from them, the symmetry and size of these conduits,— for nature doing nothing in vain, would never have given them so large a relative size without a purpose,—or from the arrangement and intimate structure of the valves in particular, and of the other parts of the heart in general, with many things besides, I frequently and seriously bethought me, and long revolved in my mind, what might be the quantity of blood which was transmitted, in how short a time its passage might be effected, and the like: and not finding it possible that this could be supplied by the juices of the ingested aliment without the veins on the one hand becoming drained, and the arteries on the other getting ruptured through the excessive charge of blood, unless the blood should somehow find its way from the arteries into the veins, and so return to the right side of the heart; I began to think whether there might not be A MOTION, AS IT WERE, IN A CIRCLE. Now this I afterwards found to be true; and I finally saw that the blood, forced by the action of the left ventricle into the arteries, was distributed to the body at large, and its several parts, in the same manner as it is sent through the lungs, impelled by the right ventricle into the pulmonary artery, and that it then passed through the veins and along the vena cava, and so round to the left ventricle in the manner already indicated. Which motion we may be allowed to call circular, in the same way as Aristotle says that the air and the rain emulate the circular motion of the superior bodies; for the moist earth, warmed by the sun, evaporates; the vapours drawn upwards are condensed, and descending in the form of rain, moisten the earth again; and by this arrangement are generations of living things produced; and in like manner too are tempests and meteors engendered by the circular motion, and by the approach and recession of the sun.

CHAPTER 5

THE PRESSURE OF THE AIR

15. From a letter of EVANGELISTA TORRICELLI to MICHEL-ANGELO RICCI

[Torricelli explains his experiment on the vacuum.]

Florence, 11 June 1644

I have already intimated to you that a certain physical experiment was being performed on the vacuum; not simply to produce a vacuum, but to make an instrument which would show the changes in the air, which is at times heavier and thicker and at times lighter and more rarefied. Many have said that a vacuum cannot be produced, others that it can be produced, but with repugnance on the part of Nature and with difficulty; so far, I know of no one who has said that it can be produced without effort and without resistance on the part of Nature. I reasoned in this way: if I were to find a plainly apparent cause for the resistance which is felt when one needs to produce a vacuum, it seems to me that it would be vain to try to attribute that action, which patently derives from some other cause, to the vacuum; indeed, I find that by making certain very easy calculations, the cause I have proposed (which is the weight of the air) should in itself have a greater effect than it does in the attempt to produce a vacuum. I say this because some Philosopher, seeing that he could not avoid the admission that the weight of the air causes the resistance which is felt in producing a vacuum, did not say that he admitted the effect of the weight of the air, but persisted in asserting that Nature also contributes at least to the abhorrence of a vacuum. We live submerged at the bottom of an ocean of the element air, which by unquestioned experiments is known to have weight, and so much, indeed, that near the surface of the earth where it is most dense, it weighs (volume for volume) about the four-hundredth part of the weight of water. Those who have written about twilight, moreover, have observed that the vaporous and visible air rises above us about fifty or fifty-four miles; I do not, however, believe its height is as great as this, since if it were, I could show that the vacuum would have to offer much greater resistance than it does—even though there is in their favour the argument that the weight referred to by Galileo applies to the air in very low places where men and animals live, whereas that on the tops of high mountains begins to be distinctly rare and of much less weight than the four-hundredth part of the weight of water.

[Torricelli now described his experiments with the apparatus shown in the Figure. A long glass tube, sealed at one end, was filled with mercury, a finger placed over the open end, and then the tube was upturned and the finger removed under mercury in a basin. The mercury stood to a height of 30 inches in the tube, having a vacuum at the top. Torricelli concluded that it was the pressure of the atmosphere that supported the mercury and continued as follows.]

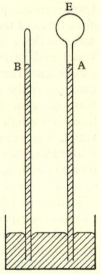

The above conclusion was confirmed by an experiment made at the same time with a vessel A and a tube B, in which the quicksilver always came to rest at the same level, AB. This is an almost certain indication that the force was not within; because if that were so, the vessel AE would have had greater force, since within it there was more rarefied material to attract the quicksilver, and a material much more powerful than that in the very small space B, on account of its greater rarefaction. I have since tried to consider from this point of view all the kinds of repulsions which are felt in the various effects attributed to vacuum, and thus far I have not encountered anything which does not go (to confirm my opinion).

16. From the account, submitted by MONSIEUR PERIER to MONSIEUR PASCAL, of the experiment performed on the Puy de Dôme, 19 September 1648

[Perier's achievement was to show that the height of the mercury in a barometer is less at the summit of a mountain than at the foot.]

The weather on Saturday last, the nineteenth of this month, was very unsettled. At about five o'clock in the morning, however, it seemed sufficiently clear; and since the summit of the Puy de Dôme was then visible, I decided to go there to make the attempt....

On that day, therefore, at eight o'clock in the morning, we started off all together for the garden of the Minim Fathers, which is almost the lowest spot in the town, and there began the experiment in this manner.

First, I poured into a vessel six pounds of quicksilver which I had rectified during the three days preceding; and having taken glass tubes of the same size, each four feet long and hermetically sealed at one end but open at the other, I placed them in the same vessel and carried out with each of them the usual vacuum experiment. Then, having set them up side by side without lifting them out of the vessel, I found that the quicksilver left in each of them stood at the same level, which was twenty-six inches[1] and three and a half lines above the surface of the quicksilver in the vessel. I repeated this experiment twice at this same spot, in the same tubes, with the same quicksilver, and in the same vessel; and found in each case that the quicksilver in the two tubes stood at the same horizontal level, and at the same height as in the first trial.

That done, I fixed one of the tubes permanently in its vessel for continuous experiment. I marked on the glass the height of the quicksilver, and leaving that tube where it stood, I requested Revd. Father Chastin, one of the brothers of the house, a man as pious as he is capable, and one who reasons very well upon these matters, to be so good as to observe from time to time all day any changes that might occur. With the other tube and a portion of the same quicksilver, I then proceeded with all these gentlemen to the top of the Puy de Dôme, some 500 fathoms above the Convent. There, after I had made the same experiments in the same way that I had made them at the Minims, we found

---

[1] A French inch equals 1·065 English inches.

that there remained in the tube a height of only twenty-three inches and two lines of quicksilver; whereas in the same tube, at the Minims we had found a height of twenty-six inches and three and a half lines. Thus between the heights of the quicksilver in the two experiments there proved to be a difference of three inches one line and a half. We were so carried away with wonder and delight, and our surprise was so great that we wished, for our own satisfaction, to repeat the experiment. So I carried it out with the greatest care five times more at different points on the summit of the mountain, once in the shelter of the little chapel that stands there, once in the open, once shielded from the wind, once in the wind, once in fine weather, once in the rain and fog which visited us occasionally. Each time I most carefully rid the tube of air; and in all these experiments we invariably found the same height of quicksilver. This was twenty-three inches and two lines, which yields the same discrepancy of three inches, one line and a half in comparison with the twenty-six inches, three lines and a half which had been found at the Minims. This satisfied us fully.

Later, on the way down at a spot called Lafon de L'Arbre, far above the Minims but much farther below the top of the mountain, I repeated the same experiment, still with the same tube, the same quicksilver, and the same vessel, and there found that the height of the quicksilver left in the tube was twenty-five inches. I repeated it a second time at the same spot; and Monsieur Mosnier, one of those previously mentioned, having the curiosity to perform it himself, then did so again, at the same spot. All these experiments yielded the same height of twenty-five inches, which is one inch, three lines and a half less than that which we had found at the Minims, and one inch and ten lines more than we had just found at the top of the Puy de Dôme. It increased our satisfaction not a little to observe in this way that the height of the quicksilver diminished with the altitude of the site.

On my return to the Minims I found that the (quicksilver in the) vessel I had left there in continuous operation was at the same height at which I had left it, that is, at twenty-six inches, three lines and a half; and the Revd. Father Chastin, who had remained there as observer, reported to us that no change had occurred during the whole day, although the weather had been very unsettled, now clear and still, now rainy, now very foggy and now windy.

Here I repeated the experiment with the tube I had carried to the Puy de Dôme, but in the vessel in which the tube used for the continuous experiment was standing. I found that the quicksilver was at the same level in both tubes, and exactly at the height of twenty-six inches three lines and a half, at which it had stood that morning in this same tube, and as it had stood all day in the tube used for the continuous experiment.

I repeated it again a last time, not only in the same tube I had used on the Puy de Dôme, but also with the same quicksilver and in the same vessel that I had carried up the mountain; and again I found the quicksilver at the same height of twenty-six inches, three lines and a half which I had observed in the morning, and thus finally verified the certainty of our results.

17. From OTTO VON GUERICKE'S *New Magdeburg Experiments on the Vacuum* (1672)

[Von Guericke records some of the difficulties he encountered in trying to create a vacuum.]

I filled a cask with water, made it everywhere air-tight, connected it on the lower side with a metal pump wherewith to draw out the water; I reasoned that as I drew out the water the part of the cask above the water would then be 'empty space'. At the first experiment the cask flew to pieces; I then affixed heavier screws, three men succeeded in pumping out the water, but then a sizzling sound was heard, the air filled the space from which water was drawn. Then I tried putting a smaller cask within the larger, so as to avoid the air rushing into the 'empty space', and drawing thence the water. Again the sound, now like the twittering of a bird, was heard and lasted for three days, for the wood was porous and let the air through. Therefore I tried a copper sphere instead; first this burst with a loud report. I attributed this to a probable defect in the spherical shape. With the greatest care I had a perfect sphere constructed. Now finally a vacuum was obtained; opening the cock attached to the sphere, the air rushed in with great violence.

18. From ROBERT BOYLE'S *A Continuation of New Experiments touching the Spring and Weight of Air* (1669)

EXPERIMENT XV

About the greatest height to which water can be raised
by Attraction or Sucking Pumps

Having met with an opportunity to borrow a place somewhat convenient to make a Tryal to what height Water may be rais'd by Pumping, I thought fit not to neglect it....

Wherefore, partly because a Tryal of such moment seem'd not to have yet been duely made by any; and partly because the varying weight of the Atmosphere was not (that appears) known, nor (consequently) taken into consideration by the ingenious Monsieur Paschal in his famous Experiment, which yet is but analogous to this; and partly because some very Late as well as Learned Writers have not acquiesc'd in his experiment, but do adhere to the old Doctrine of the Schools, which would have Water raiseable in Pumps to any height, *ob fugam vacui,* (as they speak,) I thought fit to make the best shift I could to make the Tryal, of which I now proceed to give Your Lordship an Account.

The place I borrowed for this purpose was a flat Roof about 30 foot high from the ground, and with Railes along the edges of it. The Tube we made use of should have been of Glass, if we could have procured one long and strong enough. But that being exceeding difficult, especially for me, who was not near a Glass-house, we were fain to cause a Tin-man to make several Pipes of above an inch bore, (for of a great length 'twas alleadg'd they could not be made slenderer,) and as long as he could, of Tin or Laton, as they call thin Plates of Iron Tinn'd over; and these being very carefully soder'd together made up one Pipe, of about one or two and thirty foot long, which being tied to a Pole we tried with Water whether it were stanch, and by the effluxions of that Liquor finding where the Leaks were, we caus'd them to be stopt with Soder, and then for greater security the whole Pipe, especially at the Commissures, was diligently cas'd over with our close black Cement, upon which Plaister of Paris was strewed to keep it from sticking to their hands or cloaths that should manage the Pipe. At the upper part of which was very carefully fastned with the like Cement a strong Pipe of Glass, of between 2 and

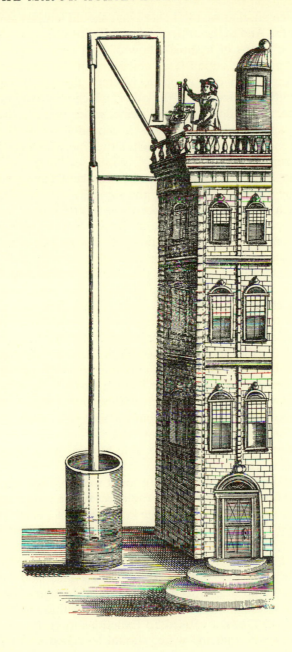

3 foot in length, that we might see what should happen at the top of the water. And to the upper part of this Pipe was (with Cement, and by the means of a short elbow of Tin) very closely fastned another Pipe of the same Metal, consisting of two pieces, making a right Angle with one another, whereof the upper part was parallel to the Horizon, and the other, which was parallel to the Glass-pipe, reacht down to the Engine, which was plac'd on the flat Roof, and was to be with good Cement sollicitously fastned to the lower end of this descending part of the Pipe, whose Horizontal leg was supported by a piece of Wood, nail'd to the above mentioned Railes; as the Tube also was kept from overmuch shaking by a board, (fasten'd to the same Railes,) and having a deep Notch cut in it, for the Tube to be inserted into.

This Apparatus being made, and the whole Tube with its Pole erected along the Wall, and fastned with strings and other helps, and the descending Pipe being carefully cemented on to the Engine, there was plac'd under the bottom of the long Tube a convenient vessel, whereinto so much Water was poured, as reach'd a great way above the orifice of the Pipe, and one was appointed to stand by to pour in more as need should require, that the vessel might be still kept competently full.

After all this the Pump was set on work, but when the water had been raised to a great height, and consequently had a great Pressure against the sides of the Tube, a small Leak or two was either discovered or made, which without moving the Tube we caus'd to be well stoppt, by one that was sent up a Ladder to apply store of Cement where it was requisite.

Wherefore at length we were able after a pretty number of Exuctions, to raise the Water to the middle of the Glass-pipe, above mentioned, but not without great store of bubbles, (made by the Air formerly conceal'd in the pores of the water, and now emerging,) which for a pretty while kept a kind of Foam upon the surface of it, (fresh ones continually succeeding those that broke.) And finding the Engine and Tube as stanch as could be well expected, I thought it a fit season to trie what was the utmost height to which Water could by Suction be elevated, and therefore though the Pump seem'd to have been plyed enough already, yet for further satisfaction, when the Water was within few inches of the top of the Glass, I caused 20 Exuctions more to be nimbly made; to be sure that the water should be raised as high as by our Pump it could be possibly. And having taken notice where the Surface rested, and caus'd a piece of Cement to be stuck near it, (for we could not then come to reach it exactly,) and descending to the Ground where the stagnant water stood, we caus'd a string to be let down, with a weight hanging at the end of it, which we applied to a mark, that had been purposely made at that part of the (Metalline) Tube, which the superficies of the stagnant water had rested at, when the water was elevated to its full height; and the other end of the string being, by him that let it down, applied to that part of the Glass as near as he could guess, where the upper part of the Water reacht, the Weight was pull'd up; and the length of the string, and (consequently) the height of the Cylinder of Water was measur'd, which amounted to 33 foot, and about 6 inches. Which done, I return'd to my lodging, which was not far off, to look upon the Baroscope, to be informed of the present weight of the Atmosphere, which I found to be but moderate, the Quick-silver standing at 29 inches, and between 2 and 3 eights of an inch. This being taken notice of, it was not difficult to compare the success of the Experiment with our Hypothesis. For if we suppose the most received proportion in bulk between Cylinders of Quick-silver and of Water of the same weight, namely that of 1 to 14, the height of the Water ought to have been 34 foot and about two inches, which is about 8 inches greater than we found it.

CHAPTER 6

THE EARLY MICROSCOPISTS

19. From ROBERT HOOKE'S *Micrographia* (1665)

[In this book Hooke records observations with his microscope of a variety of objects, such as the point of a pin, the edge of a razor, the pores of cork, insects, etc.]

OBSERVATION LIII

Of a Flea

The strength and beauty of this small creature, had it no other relation at all to man, would deserve a description.

For its strength, the *Microscope* is able to make no greater discoveries of it than the naked eye, but onely the curious contrivance of its leggs and joints, for the exerting that strength, is very plainly manifested, such as no other creature, I have yet observ'd, has any thing like it; for the joints of it are so adapted, that he can, as 'twere, fold them short one within another, and suddenly stretch, or spring them out to their whole length, that is, of the fore-leggs, the part *A* [see Figure] lies within *B*, and *B* within *C*, parallel to, or side by side each other; but the parts of the two next, lie quite contrary, that is, *D* without *E*, and *E* without *F*, but parallel also; but the parts of the hinder leggs, *G*, *H*, and *I*, bend one within another, like the parts of a double jointed Ruler, or like the foot, legg and thigh of a man; these six legs he clitches up altogether, and when he leaps, springs them all out, and thereby exerts his whole strength at once.

But, as for the beauty of it, the *Microscope* manifest it to be all over adorn'd with a curiously polish'd suit of sable Armour, neatly jointed, and beset with multitudes of sharp pinns, shaped almost like a Porcupine's Quills, or bright conical steel-bodkins; the head is on either side beautify'd with a quick and round black eye *K*, behind each of which also appears a small cavity, *L*, in which he seems to move to and fro a certain thin film beset with

36

many small transparent hairs, which probably may be his ears; in the forepart of his head, between the two fore-leggs, he has two small long jointed feelers, or rather smellers, *MM* which have four joints, and are hairy, like those of several other creatures; between these it has a small *proboscis*, or *probe*, *NNO*, that seems to consist of a tube *NN*, and a tongue or sucker *O*, which I have perceiv'd him to slip in and out. Besides these, it has also two chaps or biters *PP*, which are somewhat like those of an Ant, but I could not perceive them tooth'd; these were shaped very like the blades of a pair of round top'd Scizers, and were opened and shut just after the same manner; with these instruments does this little busie Creature bite and pierce the skin, and suck out the blood of an Animal, leaving the skin inflamed with a small round red spot. These parts are very difficult to be discovered, because, for the most part, they lye covered between the fore-legs. There are many other particulars, which, being more obvious, and affording no great matter of information, I shall pass by, and refer the Reader to the Figure.

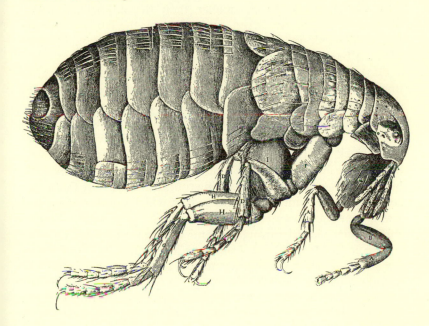

20. From ANTONY VAN LEEUWENHOEK's letters to the Royal Society on his 'Little Animals' (1676–92)

[In these letters Leeuwenhoek describes, with his characteristic, racy ingenuousness, his pioneer explorations of the world of micro-organisms. The first letter recounts the discovery of the protozoa in rain water and the attempts of *Vorticella* to 'get its tail loose' (*Vorticella* anchors itself by its stalk). The second letter is one of several in which Leeuwenhoek stoutly maintains, in face of scepticism, that a drop of water may contain millions of micro-organisms. The third letter deals with the teeming and lively bacteria found in the human mouth.]

From *Letter* 18, 9 October 1676

In the year 1675, about half-way through September (being busy with studying air, when I had much compressed it by means of water), I discovered living creatures in rain, which had stood but a few days in a new tub, that was painted blue within. This observation provoked me to investigate this water more narrowly; and especially because these little animals were, to my eye, more than ten thousand times smaller than the animalcule which Swammerdam has portrayed, and called by the name of Water-flea, or Water-louse, which you can see alive and moving in water with the bare eye.

Of the first sort that I discovered in the said water, I saw, after divers observations, that the bodies consisted of 5, 6, 7, or 8 very clear globules, but without being able to discern any membrane or skin that held these globules together, or in which they were inclosed. When these animalcules bestirred 'emselves, they sometimes stuck out two little horns, which were continually moved, after the fashion of a horse's ears. The part between these little horns was flat, their body else being roundish, save only that it ran somewhat to a point at the hind end; at which pointed end it had a tail, near four times as long as the whole body, and looking as thick, when viewed through my microscope, as a spider's web. At the end of this tail there was a pellet, of the bigness of one of the globules of the body; and this tail I could not perceive to be used by them for their movements in very clear water. These little animals were the most wretched creatures that I have ever seen; for when, with the pellet, they did but hit on any particles or little filaments (of which there are many in water, especially if it hath but stood some days), they stuck intangled in them; and then pulled their body out into an oval, and did struggle, by strongly stretching themselves, to get their tail loose; whereby their whole body then sprang back towards the pellet of the tail, and their tails then coiled up serpentwise, after the fashion of a copper or iron wire that, having been wound close about a round stick, and then taken off, kept all its windings. This motion, of stretching out and pulling together the tail, continued; and I have seen several hundred animalcules, caught fast by one another in a few filaments, lying within the compass of a coarse grain of sand.

I also discovered a second sort of animalcules, whose figure was an oval; and I imagined that their head was placed at the pointed end. These were a little bit bigger than the animalcules first mentioned. Their belly is flat, provided with divers incredibly thin little feet, or little legs, which were moved very nimbly, and which I was able to discover only after sundry great efforts, and wherewith they brought off incredibly quick motions. The upper part of their body was round, and furnished inside with 8, 10, or 12 globules: otherwise these animalcules were very clear. These little animals would change their body into a perfect round, but mostly when they came to lie high and dry. Their body was also very yielding: for if they so much as brushed against a tiny filament, their body bent in, which bend also presently sprang out again; just as if you stuck your finger into a bladder full of water, and then, on removing the finger, the inpitting went away. Yet the greatest marvel was when I brought any of the animalcules on a dry place, for I then saw them change themselves at last into a round, and then the upper part of the body rose up pyramid-like, with a point jutting out in the middle; and after having thus lain moving with their feet for a little while, they burst asunder, and the globules and a watery humour flowed away on all sides, without my being able to discern even the least sign of any skin wherein these globules and the liquid had, to all appearance, been inclosed; and at such times I could discern more globules than when they were alive. This bursting asunder I figure to myself to happen thus: imagine, for example, that you have a sheep's bladder filled with shot, peas, and water; then, if you were to dash it apieces on the ground, the shot, peas, and water would scatter themselves all over the place. . . .

From *Letter* 19, 23 March 1677

Your very welcome letters of the 12th and 22nd *ultimo* have reached me safely. I was glad to see that Mr Boyle and Mr Grew sent me their remembrances: please give these gentlemen, on my behalf, my most respectful greetings. 'Twas also a pleasure to me to see that the other Philosophers liked my observations on water, etc., though they found it hard to conceive of the huge number of little animals present in even a single drop of water. Yet I can't wonder at it, since 'tis difficult to comprehend such things without getting a sight of 'em.

But I have never affirmed, that the animals in water were present in such-and-such a number: I always say, that I imagine I see so many.

My division of the water, and my counting of the animalcules, are done after this fashion. I suppose that a drop of water doth equal a green pea in bigness; and I take a very small quantity of water, which I cause to take on a round figure, of very near the same size as a millet-seed. This latter quantity of water I figure to myself to be the one-hundredth part of the foresaid drop: for I reckon that if the diameter of a millet-seed be taken as 1, then the diameter of a green pea must be quite $4\frac{1}{2}$. This being so, then a quantity of water of the bigness of a millet-seed maketh very nearly the $\frac{1}{91}$ part of a drop, according to the received rules of mathematicks....

This amount of water, as big as a millet-seed, I introduce into a clean little glass tube (whenever I wish to let some curious person or other look at it). This slender little glass tube, containing the water, I divide again into 25 or 30, or more, parts; and I then bring it before my microscope, by means of two silver or copper springs, which I have attached thereto for this purpose, so as to be able to place the little glass tube before my microscope in any desired position, and to be able to push it up or down according as I think fit.

I showed the foresaid animalcules to a certain Gentleman, among others, in the manner just described; and he judged that he saw, in the $\frac{1}{30}$th part of a quantity of water as big as a millet-seed, more than 1000 living creatures. This same Gentleman beheld this sight with great wonder, and all the more because I told him that in this very water there were yet 2 or 3 sorts of even much smaller creatures that were not revealed to his eyes, but which I could see by means of other glasses and a different method (which I keep for myself alone). Now supposing that this Gentleman really saw 1000 animalcules in a particle of water but $\frac{1}{30}$th of the bigness of a millet-seed, that would be 30,000 living creatures in a quantity of water as big as a millet-seed, and consequently 2,730,000 living creatures in one drop of water....

From *Letter* 75, 16 September 1692

In my letter of the 12th of September, 1683, I spake, among other things, of the living creatures that are in the white matter which lieth, or groweth, betwixt or upon one's front teeth or one's grinders. Since that time, and especially in the last two or three years, I have examined this stuff divers times; but to my surprise, I could discern no living creatures in it.

Being unable to satisfy myself about this, I made up my mind to put my back into the job, and to look into the question as carefully as I could. But because I keep my teeth uncommon clean, rubbing them with salt every morning, and after meals generally picking them with a fowl's quill, or pen; I therefore found very little of the said stuff stuck on the outside of my front teeth: and in what I got out from between them, I could find nothing with life in it. Thereupon I took a little of the stuff that was on my frontmost grinders; but though I had two or three shots at these observations, 'twas not till the third attempt that I saw one or two live animalcules. Yet I could well make out some particles lying about that I felt sure must have been animalcules. This put me in a quandary again, seeing that at and about the time when I wrote to you concerning these animalcules, I never failed to see there was life in them: but though now I used just the very same magnifying-glass and apparatus (which I judged to be that best suited to the purpose), yet I couldn't make out any living creatures at all.

Having allowed my speculations to run on this subject for some time, methinks I have now got to the bottom of the dying-off of these animalcules. The reason is, I mostly or pretty near always of a morning drink coffee, as hot as I can, so hot that it puts me into a sweat: beyond this I seldom drink anything save at mealtimes in the middle of the day and in the evening; and by doing so, I find myself in the best of health. Now the animalcules that are in the white matter on the front-teeth, and on the foremost of the back teeth, being unable to bear the hotness of the coffee, are thereby killed: like I've often shown that the animalcules which are in water are made to die by a slight heating.

Accordingly, I took (with the help of a magnifying mirror) the stuff from off and from between the teeth further back in my mouth, where the heat of the coffee couldn't get at it. This stuff I mixt with a little spit out of my mouth (in which there were no air-bubbles), and I did all this in the way I've always done: and then I saw with as great a wonderment as ever before, an unconceivably great number of little animalcules, and in so unbelievably small a quantity of the foresaid stuff, that those who didn't see it with their own eyes could scarce credit it. These animalcules, or most all of them, moved so nimbly among one another, that the whole stuff seemed alive and a-moving.

CHAPTER 7

THE SEVENTEENTH CENTURY

21. From *New Atlantis* by FRANCIS BACON, Lord Verulam
(1627)

[The writer of the tale sets sail from Peru for China and Japan by the South
Sea, and after many vicissitudes comes upon the island of New Atlantis.
Here he is shown an institute of scientific research known as Salomon's
House. The first part of this extract, an abridged account of the research
being conducted there, is a remarkable prevision of modern applied
science.

The second part of the extract reveals Bacon's weakness in his conception
of scientific method.]

The End of our Foundation is the Knowledge of Causes, and
Secrett Motions of Things; And the Enlarging of the bounds of
Humane Empire, to the Effecting of all Things possible.

The Preparations and Instruments are these. We have large and
deepe Caves of severall Depths. These Caves we call the Lower
Region; And wee use them for all Coagulations, Indurations,
Refrigerations, and Conservations of Bodies. We use them likewise for the Imitation of Naturall Mines; And the Producing also
of New Artificiall Mettalls, by Compositions and Materialls which
we use, and lay ther for many yeares. Wee use them also sometimes,
(which may seeme strange,) for Curing of some Diseases, and for
Prolongation of Life.

We have High Towers; The Highest about halfe a Mile in
Height. Wee use these Towers, according to their severall
Heights, and Situations, for Insolation, Refrigeration, Conserva-
tion; And for the View of divers Meteors; As Windes, Raine, Snow,
Haile; And some of the Fiery Meteors also.

We have great Lakes, both Salt, and Fresh; wherof we have use
for the Fish, and Fowle.

We have also a Number of Artificiall Wels, and Fountaines,
made in Imitation of the Naturall Sources and Baths.

We have also Great and Spatious Houses, wher wee imitate and demonstrate Meteors.

We have also certaine Chambers, which wee call Chambers of Health, wher wee qualifie the Aire as we thinke good and proper for the Cure of diverse Diseases, and Preservation of Health.

We have also large and various Orchards and Gardens; Wherin we do not so much respect Beauty, as Variety of Ground and Soyle, proper for diverse Trees, and Herbs. And we make (by Art) in the same Orchards and Gardens, Trees and Flowers, to come earlier, or later, then their Seasons; And to come up and beare more speedily then by their Naturall Course they doe. We make them also by Art greater much then their Nature; And their Fruit greater, and sweeter, and of differing Tast, Smell, Colour, and Figure, from their Nature.

We have also Parks, and Enclosures of all Sorts, of Beasts, and Birds; which wee use not onely for View or Rarenesse, but likewise for Dissections and Trialls; That thereby we may take light, what may be wrought upon the Body of Man. We try also all Poysons, and other Medicines upon them, as well of Chyrurgery as Phisicke. By Art likewise, we make them Greater or Taller than their Kinde is; And contrary-wise Dwarfe them and stay their Grouth; Wee make them more Fruitfull and Bearing then their Kind is; and contrary-wise Barren and not Generative. Also we make them differ in Colour, Shape, Activity, many wayes.

Wee have also Particular Pooles, wher we make Trialls upon Fishes, as we have said before of Beasts, and Birds.

Wee have also Places for Breed and Generation of those Kindes of Wormes, and Flies, which are of Speciall Use; Such as are with you your Silkwormes, and Bees.

I will not hold you long with recounting of our Brew-Howses, Bake-Howses, and Kitchins, wher are made diverse Drinks, Breads, and Meats, Rare, and of speciall Effects.

Wee have also diverse Mechanicall Arts, which you have not; And Stuffes made by them; As Papers, Linnen, Silks, Tissues; dainty Works of Feathers of wonderfull Lustre; excellent Dies, and many others.

Wee have also Fournaces of great Diversities, and that keepe great Diversitie of Heates. These diverse Heates wee use, As the Nature of the Operation, which wee intend, requireth.

Wee have also Perspective-Houses, wher wee make Demonstrations of all Lights, and Radiations; And of all Colours. Wee procure meanes of Seeing Objects a-farr off. Wee have also Glasses and Meanes, to see Small and Minute Bodies, perfectly and distinctly; As the Shapes and Colours of Small Flies and Wormes, Graines and Flawes in Gemmes which cannot otherwise be seen, Observations in Urine and Bloud not otherwise to be seen.

Wee have also Sound-Houses, wher wee practise and demonstrate all Sounds and their Generation. Wee have certaine Helps, which sett to the Eare doe further the Hearing greatly. Wee have also meanes to convey Sounds in Trunks and Pipes, in strange Lines, and Distances.

Wee have also Perfume-Houses; wherwith we joyne also Practises of Tast. Wee Multiply Smells, which may seeme strange. Wee make diverse Imitations of Tast likewise, so that they will deceyve any Mans Tast.

Wee have also Engine-Houses, wher are prepared Engines and Instruments for all Sorts of Motions. Wee imitate also Flights of Birds; Wee have some Degrees of Flying in the Ayre. Wee have Shipps and Boates for Going under Water, and Brooking of Seas.

Wee have also Houses of Deceits of the Senses; wher we represent all manner of Feates of Iugling, False Apparitions, Impostures and Illusions.

These are (my Sonne) the Riches of Salomons House.

*　　*　　*

For the severall Employments and Offices of our Fellowes; Wee have Twelve that Sayle into Forraine Countries, under the Names of other Nations, (for our owne wee conceale;) Who bring us the Bookes, and Abstracts, and Patternes of Experiments of all other Parts. These wee call Merchants of Light.

Wee have Three that Collect the Experiments which are in all Bookes. These wee call Depredatours.

Wee have Three that Collect the Experiments of all Mechanicall Arts; And also of Liberall Sciences; And also of Practises which are not Brought into Arts. These we call Mystery-Men.

Wee have Three that try New Experiments, such as themselves thinke good. These wee call Pioners or Miners.

Wee have Three that Drawe the Experiments of the Former Foure into Titles, and Tables, to give the better light, for the drawing of Observations and Axiomes out of them. These wee call compilers.

We have Three that bend themselves, Looking into the Experiments of their Fellowes, and cast about how to draw out of them Things of Use, and Practise for Mans life, and Knowledge, as well for Workes, as for Plaine Demonstration of Causes, Meanes of Naturall Divinations, and the easie and cleare Discovery, of the Vertues and Parts of Bodies. These wee call Dowry-men or Benefactours.

Then after diverse Meetings and Consults of our whole Number, to consider of the former Labours and Collections, wee have Three that take care, out of them, to direct New Experiments, of a Higher Light, more Penetrating into Nature then the Former. These wee call Lamps.

Wee have Three others that doe Execute the Experiments so Directed, and Report them. These wee call Inoculatours.

Lastly, wee have Three that raise the former Discoveries by Experiments, into Greater Observations, Axiomes, and Aphorismes. These wee call Interpreters of Nature.

22. From FRANCIS BACON'S *Novum Organum* (1620)

[Bacon explains his new inductive method of making generalisations or laws from the results of observations and experiments in scientific investigation.]

In establishing axioms, another form of induction must be devised than has hitherto been employed; and it must be used for proving and discovering not first principles (as they are called) only, but also the lesser axioms, and the middle, and indeed all. For the induction which proceeds by simple enumeration is childish; its conclusions are precarious, and exposed to peril from a contradictory instance; and it generally decides on too small a number of facts, and on those only which are at hand. But the induction which is to be available for the discovery and demonstration of sciences and arts, must analyse nature by proper rejections and exclusions; and then, after a sufficient number of negatives, come to a conclusion on the affirmative instances; which has not yet been done or even attempted, save only by Plato, who does indeed employ this form of induction to a certain extent for the purpose of discussing definitions and ideas. But in order to furnish this induction or demonstration well and duly for its work, very many things are to be provided which no mortal has yet thought of; insomuch that greater labour will have to be spent in it than has hitherto been spent on the syllogism. And this induction must be used not only to discover axioms, but also in the formation of notions. And it is in this induction that our chief hope lies.

23. From RENÉ DESCARTES' *Discourse on Method* (1637)

[Descartes describes the discovery of his method, which was the construction of science and philosophy by steps, proceeding from simple self-evident truths to the more complex, in the manner of geometry.]

I was in Germany at the time; the fortune of war (the war that is still going on) had called me there. While I was returning to the army from the Emperor's coronation, the onset of the winter held me up in quarters in which I found no conversation to interest me; and since, fortunately, I was not troubled by any cares or passions, I spent the whole day shut up alone in a stove-heated room, and was at full liberty to discourse with myself about my own thoughts. . . .

I thought the following four [rules] would be enough, provided that I made a firm and constant resolution not to fail even once in the observance of them.

The first was never to accept anything as true if I had not evident knowledge of its being so; that is, carefully to avoid precipitancy and prejudice, and to embrace in my judgement only what presented itself to my mind so clearly and distinctly that I had no occasion to doubt it.

The second, to divide each problem I examined into as many parts as was feasible, and as was requisite for its better solution.

The third, to direct my thoughts in an orderly way; beginning with the simplest objects, those most apt to be known, and ascending little by little, in steps as it were, to the knowledge of the most complex; and establishing an order in thought even when the objects had no natural priority one to another.

And the last, to make throughout such complete enumerations

and such general surveys that I might be sure of leaving nothing out.

Those long chains of perfectly simple and easy reasonings by means of which geometers are accustomed to carry out their most difficult demonstrations had led me to fancy that everything that can fall under human knowledge forms a similar sequence; and that so long as we avoid accepting as true what is not so, and always preserve the right order to deduction of one thing from another, there can be nothing too remote to be reached in the end, or too well hidden to be discovered.

24. From THOMAS HOBBES' *Leviathan* (1651)

[In these two brief extracts, the first taken from chapter VI and the second from chapter XV, Hobbes denies that Good is an absolute value and puts forward the view that moral philosophy is concerned with the natural laws of man's appetites and aversions.]

But whatsoever is the object of any mans Appetite or Desire; that is it, which he for his part calleth *Good*: And the object of his Hate, and Aversion, *Evill*; And of his Contempt, *Vile* and *Inconsiderable*. For these words of Good Evill, and Contemptible, are ever used with relation to the person that useth them: There being nothing simply and absolutely so; nor any common Rule of Good and Evill, to be taken from the nature of the objects themselves;....

For Morall Philosophy is nothing else but the Science of what is *Good*, and *Evill*, in the conversation, and Society of man-kind. *Good*, and *Evill*, are names that signifie our Appetites, and Aversions; which in different tempers, customes, and doctrines of men, are different: And divers men, differ not onely in their Judgement, on the senses of what is pleasant, and unpleasant to the tast, smell, hearing, touch, and sight; but also of what is conformable, or disagreeable to Reason, in the actions of common life. Nay, the same man, in divers times, differs from himselfe; and one time praiseth, that is, calleth Good, what another time he dispraiseth, and calleth Evill: From whence arise Disputes, Controversies, and at last War. And therefore so long a man is in the condition of meer Nature, (which is a condition of War,) as private Appetite is the measure of Good, and Evill: And consequently all men agree on this, that Peace is Good, and therefore also the way, or means of Peace, which (as I have shewed before) are *Justice, Gratitude, Modesty, Equity, Mercy*, & the rest of the Laws of Nature, are good; that is to say, *Morall Vertues*; and their contrarie *Vices*, Evill. Now the science of Vertue and Vice, is Morall Philosophie; and therefore the true Doctrine of the Lawes of Nature, is the true Morall Philosophie.

CHAPTER 8

THE CREATION OF MODERN CHEMISTRY

25. ANTOINE LAVOISIER's sealed note of 1 November 1772

[Lavoisier tries to secure priority for his ideas on combustion—ideas which, he found later, were put forward over a century earlier.]

About eight days ago I discovered that sulphur, in burning, far from losing weight, on the contrary gains in weight; that is to say that from a pound of sulphur one can obtain much more than a pound of vitriolic acid, making allowance for humidity of the air; it is the same with phosphorus; this increase in weight arises from an immense quantity of air that fixes itself during the combustion and combines with the vapours.

This discovery, which I have established by experiments that I look upon as decisive, has led me to think that what is observed in the combustion of sulphur and phosphorus may well take place with regard to all substances that gain in weight by combustion and calcination: and I am persuaded that the increase in weight of metallic calces is due to the same cause. Experiment has completely confirmed my conjectures: I have carried out the reduction of litharge in closed vessels, using the apparatus of Hales, and observed the liberation, at the moment the calx changed into metal, of a large quantity of air, this air having a volume a thousand times greater than the amount of litharge employed. This discovery appearing to me one of the most interesting of those that have been made since the time of Stahl, I felt compelled to secure my right in it, by depositing it in the hands of the Secretary of the *Académie*, to remain sealed until the time when I shall make my experiments known.

Paris, 1st November, 1772. Lavoisier.

26. From JOSEPH PRIESTLEY'S *Experiments and Observations on Different Kinds of Air* (1775)

[Priestley describes his experiments with 'air purer than the best common air', now known as oxygen.]

There are, I believe, very few maxims in philosophy that have laid firmer hold upon the mind, than that air, meaning atmospherical air (free from various foreign matters, which were always supposed to be dissolved, and intermixed with it) is *a simple elementary substance*, indestructible, and unalterable, at least as much so as water is supposed to be. In the course of my inquiries, I was, however, soon satisfied that atmospherical air is not an unalterable thing; for that the phlogiston with which it becomes loaded from bodies burning in it, and animals breathing it, and various other chemical processes, so far alters and depraves it, as to render it altogether unfit for inflammation, respiration, and other purposes to which it is subservient; and I had discovered that agitation in water, the process of vegetation, and probably other natural processes, by taking out the superflous phlogiston, restore it to its original purity. But I own I had no idea of the possibility of going any farther in this way, and thereby procuring air purer than the best common air.

[Priestley then describes how, by heating *mercurius calcinatus per se* (mercuric oxide) with his burning glass, he obtained a gas which he called *dephlogisticated air* (oxygen).]

On the 8th of this month I procured a mouse, and put it into a glass vessel, containing two ounce-measures of the air from mercurius calcinatus. Had it been common air, a full-grown mouse, as this was, would have lived in it about a quarter of an hour. In this air, however, my mouse lived a full half hour; and though it was taken out seemingly dead, it appeared to have been only exceedingly chilled; for, upon being held to the fire, it presently revived, and appeared not to have received any harm from the experiment.

By this I was confirmed in my conclusion, that the air extracted from mercurius calcinatus, &c. was, *at least, as good* as common air; but I did not certainly conclude that it was any *better*; because, though one mouse would live only a quarter of an hour in a given quantity of air, I knew it was not impossible but that another mouse might have lived in it half an hour; so little accuracy is there in this method of ascertaining the goodness of air: and indeed I have never had recourse to it for my own satisfaction, since the discovery of that most ready, accurate, and elegant test that nitrous air furnishes. But in this case I had a view to publishing the most generally-satisfactory account of my experiments that the nature of the thing would admit of.

This experiment with the mouse, when I had reflected upon it some time, gave me so much suspicion that the air into which I had put it was better than common air, that I was induced, the day after, to apply the test of nitrous air to a small part of that very quantity of air which the mouse had breathed so long; so that, had it been common air, I was satisfied it must have been very nearly, if not altogether, as noxious as possible, so as not to be affected by nitrous air; when, to my surprize again, I found that though it had been breathed so long, it was still better than common air. For after mixing it with nitrous air, in the usual proportion of two to one, it was diminished in the proportion of $4\frac{1}{2}$ to $3\frac{1}{2}$; that is, the nitrous air had made it two ninths less than before, and this in a very short space of time; whereas I had never found that, in the longest time, any common air was reduced more than one fifth of its bulk by any proportion of nitrous air, nor more than one fourth by any phlogistic process whatever. Thinking of this extraordinary fact upon my pillow, the next morning I put another measure of nitrous air to the same mixture and, to my utter astonishment, found that it was farther diminished to almost one half of its original quantity. I then put a third measure to it; but this did not diminish it any farther: but, however, left it one measure less than it was even after the mouse had been taken out of it.

Being now fully satisfied that this air, even after the mouse had breathed it half an hour, was much better than common air; and having a quantity of it still left, sufficient for the experiment, viz. an ounce-measure and a half, I put the mouse into it; when I observed that it seemed to feel no shock upon being put into it, evident signs of which would have been visible, if the air had not been very wholesome; but that it remained perfectly at its ease another full half hour, when I took it out quite lively and vigorous. . . .

My reader will not wonder, that, after having ascertained the superior goodness of dephlogisticated air by mice living in it, and the other tests above mentioned, I should have the curiosity to taste it myself. I have gratified that curiosity, by breathing it, drawing it through a glass-syphon, and, by this means, I reduced a large jar full of it to the standard of common air. The feeling of it to my lungs was not sensibly different from that of common air; but I fancied that my breast felt peculiarly light and easy for some time afterwards. Who can tell but that, in time, this pure air may become a fashionable article in luxury. Hitherto only two mice and myself have had the privilege of breathing it.

27. From HENRY CAVENDISH'S *Experiments on Air* (1781)

[Cavendish describes how, by exploding dephlogisticated air (oxygen) and inflammable air (hydrogen), he obtained water. In the following passages he discusses the possible interpretations of the experiments.]

All the foregoing experiments, on the explosion of inflammable air with common and dephlogisticated airs, except those which relate to the cause of the acid found in the water, were made in the summer of the year 1781, and were mentioned by me to Dr PRIESTLEY, who in consequence of it made some experiments of the same kind, as he relates in a paper printed in the preceding volume of the Transactions. During the last summer also, a friend of mine gave some account of them to M. LAVOISIER, as well as of the conclusion drawn from them, that dephlogisticated air is only water deprived of phlogiston; but at that time so far was M. LAVOISIER from thinking any such opinion warranted, that, till he was prevailed upon to repeat the experiment himself, he found some difficulty in believing that nearly the whole of the two airs could be converted into water. It is remarkable, that neither of these gentlemen found any acid in the water produced by the combustion; which might proceed from the latter having burnt the two airs in a different manner from what I did; and from the former having used a different kind of inflammable air, namely, that from charcoal [i.e. carbon monoxide], and perhaps having used a greater proportion of it....

From what has been said there seems the utmost reason to think, that dephlogisticated air is only water deprived of its phlogiston, and that inflammable air, as was before said, is either phlogisticated water, or else pure phlogiston; but in all probability the former....

There are several memoirs of M. LAVOISIER published by the Academy of Sciences, in which he intirely discards phlogiston, and explains those phaenomena which have been usually attributed to the loss or attraction of that substance, by the absorption or expulsion of dephlogisticated air....

It seems therefore, from what has been said, as if the phaenomena of nature might be explained very well on this principle, without the help of phlogiston; and indeed, as adding dephlogisticated air to a body comes to the same thing as depriving it of its phlogiston and adding water to it, and as there are, perhaps, no bodies entirely destitute of water, and as I know no way by which phlogiston can be transferred from one body to another, without leaving it uncertain whether water is not at the same time transferred, it will be very difficult to determine by experiment which of these opinions is the truest; but as the commonly received principle of phlogiston explains all phaenomena, at least as well as M. LAVOISIER'S, I have adhered to that.

28. From ANTOINE LAVOISIER'S *Elementary Treatise on Chemistry* (1789)

[Lavoisier's main purpose in writing his textbook was to disseminate a new chemical nomenclature, drawn up by himself, de Morveau, Berthollet and de Fourcroy.

The following passage illustrates the new nomenclature in relation to his theory of combustion. Lavoisier explains combustion, as we do today, in terms of the chemical combination of the burning substance with oxygen. The heat liberated he calls caloric, and he regards this as released from association with oxygen.]

CHAP. VII

Of the decomposition of oxygen gas by means of metals,
and the formation of metallic oxyds

Oxygen has a stronger affinity with metals heated to a certain degree than with caloric; in consequence of which, all metallic bodies, excepting gold, silver, and platina, have the property of decomposing oxygen gas, by attracting its base from the caloric with which it was combined. We have already shown in what manner this decomposition takes place, by means of mercury and iron; having observed, that, in the case of the first, it must be considered as a kind of gradual combustion, whilst, in the latter, the combustion is extremely rapid, and attended with a brilliant flame. The use of the heat employed in these operations is to separate the particles of the metal from each other, and to diminish their attraction of cohesion or aggregation, or, what is the same thing, their mutual attraction for each other.

The absolute weight of metallic substances is augmented in proportion to the quantity of oxygen they absorb; they, at the same time, lose their metallic splendour, and are reduced into an earthy pulverulent matter. In this state metals must not be considered as entirely saturated with oxygen, because their action upon this element is counterbalanced by the power of affinity between it and caloric. During the calcination of metals, the oxygen is therefore acted upon by two separate and opposite powers, that of its attraction for caloric, and that exerted by the metal, and only tends to unite with the latter in consequence of the excess of the latter over the former, which is, in general, very inconsiderable. Wherefore, when metallic substances are oxygenated in atmospheric air, or in oxygen gas, they are not converted into acids like sulphur, phosphorus, and charcoal, but are only changed into intermediate substances, which, though approaching to the nature of salts, have not acquired all the saline properties. The old chemists have affixed the name of *calx* not only to metals in this state, but to every body which has been long exposed to the action of fire without being melted. They have converted this word calx into a generical term, under which they confound calcareous earth, which, from a neutral salt, which it really was before calcination, has been changed by fire into an earthy alkali, by *losing* half of its weight, with metals which, by the same means, have joined themselves to a new substance, whose quantity often *exceeds* half their weight, and by which they have been changed almost into the nature of acids. This mode of classifying substances of so very opposite natures, under the same generic name, would have been quite contrary to our principles of nomenclature, especially as, by retaining the above term for this state of metallic substances, we must have conveyed very false ideas of its nature. We have, therefore, laid aside the expression *metallic calx* altogether, and have substituted in its place the term *oxyd*, from the Greek word ὀξύς.

By this may be seen, that the language we have adopted is both copious and expressive. The first or lowest degree of oxygenation in bodies, converts them into oxyds; a second degree of additional oxygenation constitutes the class of acids, of which the specific names, drawn from their particular bases, terminate in *ous*, as the *nitrous* and *sulphurous* acids; the third degree of oxygenation changes these into the species of acids distinguished by the termination in *ic*, as the *nitric* and *sulphuric* acids; and, lastly, we can express a fourth, or highest degree of oxygenation, by adding the word *oxygenated* to the name of the acid, as has been already done with the *oxygenated muriatic* acid.

We have not confined the term *oxyd* to expressing the combinations of metals with oxygen, but have extended it to signify that first degree of oxygenation in all bodies, which, without converting them into acids, causes them to approach to the nature of salts. Thus, we give the name of *oxyd of sulphur* to that soft substance into which sulphur is converted by incipient combustion; and we call the yellow matter left by phosphorus, after combustion, by the name of *oxyd of phosphorus*. In the same manner, nitrous gas, which is azote in its first degree of oxygenation, is the *oxyd of azote*. We have likewise oxyds in great numbers from the vegetable and animal kingdoms; and I shall show, in the sequel, that this new language throws great light upon all the operations of art and nature.[1]

---

[1] For Lavoisier, nitrous and sulphurous acids were N_2O_3 and SO_2, nitric and sulphuric acids were N_2O_5 and SO_3. We now regard the combination of these oxides with water as the acids, namely HNO_2 and H_2SO_3, and HNO_3 and H_2SO_4 respectively. We do not now use the word oxygenated in the sense Lavoisier recommends and we use the word nitrogen instead of azote.

CHAPTER 9

THE HEROIC AGE OF GEOLOGY

29. From JAMES HUTTON'S *Theory of the Earth* (1795)

[The range of mountains running across the south of Scotland between the counties of Ayr and Berwick consists of almost vertical strata of a hard rock known as schist or schistus. On each side of the range the lower country is composed of horizontal strata of softer sandstone and marl. Hutton made a study of the junction between these vertical and horizontal strata, known as an unconformity, to support and illustrate his theory of interchanging sea and land.]

The river Tiviot has made a wide valley as might have been expected, in running over those horizontal strata of marly or decaying substances; and the banks of this river declining gradually are covered with gravel and soil, and show little of the solid strata of the country. This, however, is not the case with the Jed, which is to the southward of the Tiviot; that river, in many places, runs upon the horizontal strata, and undermines steep banks, which falling shows high and beautiful sections of the regular horizontal strata. The little rivulets also which fall into the Jed have hollowed out deep gallies in the land, and show the uniformity of the horizontal strata.

In this manner I was disposed to look for nothing more than what I had seen among those mineral bodies, when one day, walking in the beautiful valley above the town of Jedburgh, I was surprised with the appearance of vertical strata in the bed of the river, where I was certain that the banks were composed of horizontal strata. I was soon satisfied with regard to this phenomenon, and rejoiced at my good fortune in stumbling upon an object so interesting to the natural history of the earth, and which I had been long looking for in vain.

Here the vertical strata, similar to those that are in the bed of the Tweed, appear; and above those vertical strata, are placed the horizontal beds, which extend along the whole country.

57

The question which we would wish to have solved is this; if the vertical strata had been broken and erected under the super-incumbent horizontal strata; or if, after the vertical strata had been broken and erected, the horizontal strata had been deposited upon the vertical strata, then forming the bottom of the sea.

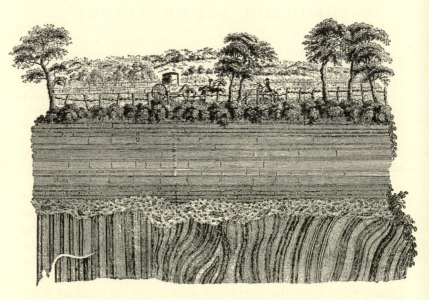

That strata, which are regular and horizontal in one place, should be found bended, broken, or disordered at another, is not uncommon; it is always found more or less in all our horizontal strata. Now, to what length this disordering operation might have been carried, among strata under others, without disturbing the order and continuity of those above, may perhaps be difficult to determine; but here, in this present case, is the greatest disturbance of the under strata, and a very great regularity among those above. Here at least is the most difficult case of this kind to conceive, if we are to suppose that the upper strata had been deposited before those below had been broken and erected.

Let us now suppose that the under strata had been disordered at the bottom of the sea, before the superincumbent bodies were deposited; it is not to be well conceived, that the vertical strata should in that case appear to be cut off abruptly, and present their regular edges immediately under the uniformly deposited substances above. But, in the case now under consideration, there appears the most uniform section of the vertical strata, their ends go up regularly to the horizontal deposited bodies. Now, in whatever state the vertical strata had been in at the time of this event, we can hardly suppose that they could have been so perfectly cut off, without any relict being left to trace that operation. It is much more probable to suppose, that the sea had washed away the relics of the broken and disordered strata, before those that are now superincumbent had been begun to be deposited. But we cannot suppose two such contrary operations in the same place, as that of carrying away the relics of those broken strata, and the depositing of sand and subtile earth in such a regular order. We are therefore led to conclude, that the bottom of the sea, or surface of those erected strata, had been in very different situations at those two periods, when the relics of the disordered strata had been carried away, and when the new materials had been deposited.

If this shall be admitted as a just view of the subject, it will be fair to suppose, that the disordered strata had been raised more or less above the surface of the ocean; that, by the effects of either rivers, winds, or tides, the surface of the vertical strata had been washed bare; and that this surface had been afterwards sunk below the influence of those destructive operations, and thus placed in a situation proper for the opposite effect, the accumulation of matter prepared and put in motion by the destroying causes.

I will not pretend to say that this has all the evidence that should be required, in order to constitute a physical truth, or principle from whence we were to reason farther in our theory; but, as a simple fact, there is more probability for the thing having happened in that manner than in any other; and perhaps this is all that may be attained, though not all that were to be wished on the occasion. Let us now see how far any confirmation may be obtained from the examination of all the attending circumstances in those operations. . . .

In describing the vertical and horizontal strata of the Jed, no mention has been made of a certain puddingstone, which is interposed between the two, lying immediately upon the one and under the other. . . . When we examine the stones and gravel of which it is composed, these appear to have belonged to the vertical strata or schistus mountains. They are in general the hard and solid parts of those indurated strata, worn and rounded by attrition; particularly sand or marl-stone, consolidated and veined with quartz, and many fragments of quartz, all rounded by attrition....

From this it will appear, that the schistus mountains or the vertical strata of indurated bodies had been formed, and had been wasted and worn in the natural operations of the globe, before the horizontal strata were begun to be deposited in those places; the gravel formed of those indurated broken bodies worn round by attrition evince that fact.

[Hutton now describes his further observations of horizontal strata overlying the vertical strata, in the valleys of the Tweed and Teviot.]

It will now be reasonable to suppose that all the schistus which we perceive, whether in the mountains or in the valleys, exposed to our view had been once covered with those horizontal strata which are observed in Berwickshire and Tiviotdale; and that, below all those horizontal strata in the level country, there is at present a body or basis of vertical or inclined schistus, on which the horizontal strata of a secondary order had been deposited. This is the conclusion that I had formed at Jedburgh, before I had seen the confirmation of it in the Tiviot; it is the only one that can be formed according to this view of things; and it must remain in the present state until more evidence be found by which the probability may be either increased or diminished....

We may now come to this general conclusion, that, in this example of horizontal and posterior strata placed upon the vertical schisti, which are prior in relation to the former, we obtain a further view into the natural history of this earth, more than what appears in the simple succession of one stratum above another. We know, in general, that all the solid parts of this earth, which come to our view, have either been formed originally by subsidence at the bottom of the sea, or been transfused in a melted state from the mineral regions among those solid bodies; but here we further learn, that the indurated and erected strata, after being broken and washed by the moving waters, had again been sunk below the sea, and had served as a bottom or basis on which to form a new structure of strata; and also, that those new or posterior strata had been indurated or cemented by the consolidating operations of the mineral region, and elevated from the bottom of the sea into the place of land, or considerably above the general surface of the waters. It is thus that we may investigate particular operations in the general progress of nature, which has for object to renovate the surface of the earth necessarily wasted in the operation of a world sustaining plants and animals.

[Hutton summarises his theory, towards the close of the book, as follows.]

Let us then take a cursory view of this system of things, upon which we have proceeded in our theory, and upon which the constitution of this world seems to depend.

Our solid earth is every where wasted, where exposed to the day. The summits of the mountains are necessarily degraded. The solid and weighty materials of those mountains are every where urged through the valleys, by the force of running water. The soil, which is produced in the destruction of the solid earth, is gradually travelled by the moving water, but is constantly supplying vegetation with its necessary aid. This travelled soil is at last deposited upon the coast, where it forms most fertile countries. But the billows of the ocean agitate the loose materials upon the shore, and wear away the coast, with the endless repetitions of this act of power, or this imparted force. Thus the continent of our earth, sapped in its foundation, is carried away into the deep, and sunk again at the bottom of the sea, from whence it had originated.

We are thus led to see a circulation in the matter of this globe, and a system of beautiful oeconomy in the works of nature. This earth, like the body of an animal, is wasted at the same time that it is repaired. It has a state of growth and augmentation; it has another state, which is that of diminution and decay. This world is thus destroyed in one part, but it is renewed in another; and the operations by which this world is thus constantly renewed, are as evident to the scientific eye, as are those in which it is necessarily destroyed. The marks of the internal fire, by which the rocks beneath the sea are hardened, and by which the land is produced above the surface of the sea, have nothing in them which is doubtful or ambiguous. The destroying operations again, though placed within the reach of our examination, and evident almost to every observer, are no more acknowledged by mankind, than is that system of renovation which philosophy alone discovers.

30. From WILLIAM SMITH'S *Stratigraphical System of Organized Fossils* (1817)

[William Smith describes how strata can be identified and traced by means of the fossils they contain.]

My original method of tracing the Strata by the organized Fossils imbedded therein, is thus reduced to a science not difficult to learn. Ever since the first written account of this discovery was circulated in 1799 it has been closely investigated by my scientific acquaintance in the vicinity of Bath; some of whom search the quarries of different Strata in that district with as much certainty of finding the characteristic Fossils of the respective rocks, as if they were on the shelves of their cabinets. By this new method of searching for organized Fossils with the regularity with which they are imbedded in such a variety of Strata, many new species have been discovered. The Geologist is thus enabled to fix the locality of those previously found; to direct the attentive investigator in his pursuits; and to find in all former cabinets and catalogues numerous proofs of accuracy in this mode of identifying the Strata.

The virtuoso will therefore now enter upon the study and selection of organized Fossils with the twofold advantage of amusement and utility. The various component parts of the soil, and all the subterraneous productions of his estate, become interesting objects of research; the contents of quarries, pits, wells, and other excavations, hitherto thought unworthy of notice, will be scrupulously examined.

The organized Fossils which may be found, will enable him to identify the Strata of his own estate with those of others: thus his lands may be drained with more certainty of success, his buildings substantially improved, and his private and public roads better made, and repaired at less expense....

Many Strata being entirely without organized Fossils, the investigation is much facilitated, by rendering the courses of those Strata which contain them more distinct; and the courses of all the Strata being known, the name of the place where any specimen is found is sufficient to mark its locality in the Strata, and the specimens being filled with the matter in which they are imbedded materially assist in identifying the Stratum to which they belong. In this respect Mineral Conchology has much the advantage of recent; the matter of the Stratum fully compensating in a geological point of view, for any defect in the specimen. Shells are generally without the animals, which are mostly incapable of preservation; fossils frequently represent the animals without the shells (i.e. the interior conformation of the shell). In general, fossil shells are so effectually closed and filled with stony matter, that the hinge, opening, and other characters, cannot be observed.

Numerous Zoophites naturally too tender for preservation, have in their fossil state their shape and most minute organization beautifully retained in limestone, flint, and other solid matter. Thus not only in clays, sands, and rocks, but in the hardest stones, are displayed all the treasures of an ancient deep, which prove the high antiquity and watery origin of the earth; for nothing can more plainly than the Zoophites evince the once fine fluidity of the stoney matter in which they are enveloped, no fluid grosser than water being capable of pervading their pores. The process which converted them and their element into stone seems to have been similar to that of freezing water, which would suddenly fix all the inhabitants of the ocean, each in its place, with all the original form and character. Organized Fossils are to the naturalist as coins to the antiquary; they are the antiquities of the earth; and very distinctly show its gradual regular formation, with the various changes of inhabitants in the watery element.

31. From GEORGES CUVIER'S *Essay on the Theory of the Earth* (1821)

[Cuvier mentions his outstanding work of reconstructing extinct animals from their fragmentary remains and then goes on to recount the evidence for his theory of geological catastrophes.]

As an antiquary of a new order, I have been obliged to learn the art of decyphering and restoring these remains, of discovering and bringing together, in their primitive arrangement, the scattered and mutilated fragments of which they are composed, of reproducing, in all their original proportions and characters, the animals to which these fragments formerly belonged, and then of comparing them with those animals which still live on the surface of the earth; an art which is almost unknown, and which presupposes, what had scarcely been obtained before, an acquaintance with those laws which regulate the coexistence of the forms by which the different parts of organized beings are distinguished. I had next to prepare myself for these inquiries by others of a far more extensive kind, respecting the animals which still exist. Nothing, except an almost complete review of creation in its present state, could give a character of demonstration to the results of my investigations into its ancient state; but that review has afforded me, at the same time, a great body of rules and affinities which are no less satisfactorily demonstrated; and the whole animal kingdom has been subjected to new laws in consequence of this Essay on a small part of the theory of the earth. . . .

When the traveller passes through those fertile plains where gently-flowing streams nourish in their course an abundant vegetation, and where the soil, inhabited by a numerous population, adorned with flourishing villages, opulent cities, and superb monuments, is never disturbed except by the ravages of war and the oppression of tyrants, he is not led to suspect that nature also has had her intestine wars, and that the surface of the globe has been much convulsed by successive revolutions and various catastrophes. But his ideas change as soon as he digs into that soil which presented such a peaceful aspect, or ascends the hills which border the plain. . . .

The lowest and most level parts of the earth, when penetrated to a very great depth, exhibit nothing but horizontal strata composed of various substances, and containing almost all of them innumerable marine productions. Similar strata, with the same kind of productions, compose the hills even to a great height. Sometimes the shells are so numerous as to constitute the entire body of the stratum. They are almost everywhere in such a perfect state of preservation, that even the smallest of them retain their most delicate parts, their sharpest ridges, and their finest and tenderest processes. They are found in elevations far above the level of every part of the ocean, and in places to which the sea could not be conveyed by any existing cause. They are not only inclosed in loose sand, but are often incrusted and penetrated on all sides by the hardest stones. Every part of the earth, every hemisphere, every continent, every island of any size, exhibits the same phenomenon. We are therefore forcibly led to believe, not only that the sea has at one period or another covered all our plains, but that it must have remained there for a long time, and in a state of tranquillity; which circumstance was necessary for the formation of deposits so extensive, so thick, in part so solid, and containing exuviae so perfectly preserved....

If we examine with greater care these remains of organized bodies, we shall discover, in the midst even of the most ancient secondary strata, other strata that are crowded with animal or vegetable productions, which belong to the land and to fresh water; and amongst the more recent strata, that is, the strata which are nearest the surface, there are some of them in which land animals are buried under heaps of marine productions. Thus the various catastrophes of our planet have not only caused the different parts of our continent to rise by degrees from the basin of the sea, but it has also frequently happened, that lands which had been laid dry have been again covered by the water, in consequence either of these lands sinking down below the level of the sea, or of the sea being raised above the level of the lands. The particular portions of the earth also which the sea has abandoned by its last retreat, had been laid dry once before, and had at that time produced quadrupeds, birds, plants, and all kinds of terrestrial productions; it had then been inundated by the sea, which has since retired from it, and left it to be occupied by its own proper inhabitants.

32. From CHARLES LYELL'S *Principles of Geology* (1830-3)

[Lyell discusses his theory of uniformitarianism, as opposed to theories of catastrophes.]

If we reflect on the history of the progress of geology, as explained in the preceding chapters, we perceive that there have been great fluctuations of opinion respecting the nature of the causes to which all former changes of the earth's surface are referable. The first observers conceived the monuments which the geologist endeavours to decipher to relate to an original state of the earth, or to a period when there were causes in activity, distinct, in kind and degree, from those now constituting the economy of nature. These views were gradually modified, and some of them entirely abandoned in proportion as observations were multiplied, and the signs of former mutations were skilfully interpreted. Many appearances, which had for a long time been regarded as indicating mysterious and extraordinary agency, were finally recognised as the necessary result of the laws now governing the material world; and the discovery of this unlooked-for conformity has at length induced some philosophers to infer, that, during the ages contemplated in geology, there has never been any interruption to the agency of the same uniform laws of change. The same assemblage of general causes, they conceive, may have been sufficient to produce, by their various combinations, the endless diversity of effects, of which the shell of the earth has preserved the memorials; and, consistently with these principles, the recurrence of analogous changes is expected by them in time to come.

Whether we coincide or not in this doctrine, we must admit that the gradual progress of opinion concerning the succession of phenomena in very remote eras, resembles, in a singular manner, that which has accompanied the growing intelligence of every people, in regard to the economy of nature in their own times. In an early state of advancement, when a greater number of natural appearances are unintelligible, an eclipse, an earthquake, a flood, or the approach of a comet, with many other occurrences afterwards found to belong to the regular course of events, are regarded as prodigies. The same delusion prevails as to moral phenomena, and many of these are ascribed to the intervention of demons, ghosts, witches, and other immaterial and supernatural agents. By degrees, many of the enigmas of the moral and physical world are explained, and, instead of being due to extrinsic and irregular causes, they are found to depend on fixed and invariable laws. The philosopher at last becomes convinced of the undeviating uniformity of secondary causes; and, guided by his faith in this principle, he determines the probability of accounts transmitted to him of former occurrences, and often rejects the fabulous tales of former times, on the ground of their being irreconcilable with the experience of more enlightened ages.

Prepossessions in regard to the duration of past time.—As a belief in the want of conformity in the causes by which the earth's crust has been modified in ancient and modern periods was, for a long time, universally prevalent, and that, too, amongst men who were convinced that the order of nature had been uniform for the last several thousand years, every circumstance which could have influenced their minds and given an undue bias to their opinions deserves particular attention. Now the reader may easily satisfy himself, that, however undeviating the course of nature may have been from the earliest epochs, it was impossible for the first cultivators of geology to come to such a conclusion, so long as they were under a delusion as to the age of the world, and the date of the first creation of animate beings. However fantastical some theories of the sixteenth century may now appear to us,— however unworthy of men of great talent and sound judgment,— we may rest assured that, if the same misconception now prevailed in regard to the memorials of human transactions, it would give rise to a similar train of absurdities. Let us imagine, for example, that Champollion, and the French and Tuscan literati when engaged in exploring the antiquities of Egypt, had visited that country with a firm belief that the banks of the Nile were never peopled by the human race before the beginning of the nineteenth century, and that their faith in this dogma was as difficult to shake as the opinion of our ancestors, that the earth was never the abode of living beings until the creation of the present continents, and of the species now existing,—it is easy to perceive what extravagant systems they would frame, while under the influence of this delusion, to account for the monuments discovered in Egypt. The sight of the pyramids, obelisks, colossal statues, and ruined temples, would fill them with such astonishment, that for a time they would be as men spell-bound—wholly incapable of reasoning with sobriety.

CHAPTER 10

THE EIGHTEENTH CENTURY

33. From JOHN LOCKE'S *An Essay concerning Human Understanding* (1690)

[Locke states the basis of his empiricism, that the sole sources of our knowledge are ideas of sensation and ideas of reflection.]

2. Let us then suppose the mind to be, as we say, white paper void of all characters, without any ideas. How comes it to be furnished? Whence comes it by that vast store which the busy and boundless fancy of man has painted on it with an almost endless variety? Whence has it all the *materials* of reason and knowledge? To this I answer, in one word, from EXPERIENCE. In that all our knowledge is founded; and from that it ultimately derives itself. Our observation, employed either about *external sensible objects, or about the internal operations of our minds perceived and reflected on by ourselves, is that which supplies our understandings with all the materials of thinking.* These two are the fountains of knowledge, from whence all the ideas we have, or can naturally have, do spring.

3. First, our Senses, conversant about particular sensible objects, do convey into the mind several distinct perceptions of things, according to those various ways wherein those objects do affect them. And thus we come by those *ideas* we have of *yellow, white, heat, cold, soft, hard, bitter, sweet,* and all those which we call sensible qualities; which when I say the senses convey into the mind, I mean, they from external objects convey into the mind what produces there those perceptions. This great source of most of the ideas we have, depending wholly upon our senses, and derived by them to the understanding, I call SENSATION.

4. Secondly, the other fountain from which experience furnisheth the understanding with ideas is the perception of the operations of our own mind within us, as it is employed about the ideas it has got; which operations, when the soul comes to reflect

on and consider, do furnish the understanding with another set of ideas, which could not be had from things without. And such are *perception, thinking, doubting, believing, reasoning, knowing, willing*, and all the different actings of our own minds; which we being conscious of, and observing in ourselves, do from these receive into our understandings as distinct ideas as we do from bodies affecting our senses. This source of ideas every man has wholly in himself; and though it be not sense, as having nothing to do with external objects, yet it is very like it, and might properly enough be called *internal sense*. But as I call the other Sensation, so I call this REFLECTION, the ideas it affords being such only as the mind gets by reflecting on its own operations within itself. By reflection then, in the following part of this discourse, I would be understood to mean, that notice which the mind takes of its own operations, and the manner of them, by reason whereof there come to be ideas of these operations in the understanding. These two, I say, viz. external material things, as the objects of SENSATION, and the operations of our own minds within, as the objects of REFLECTION, are to me the only originals from whence all our ideas take their beginnings. The term *operations* here I use in a large sense, as comprehending not barely the actions of the mind about its ideas, but some sort of passions arising sometimes from them, such as is the satisfaction or uneasiness arising from any thought.

34. From GEORGE BERKELEY'S *The Principles of Human Knowledge* (1710)

[Berkeley argues that the external world cannot exist without a mind.]

6. Some truths there are so near and obvious to the mind that a man need only open his eyes to see 'em. Such I take this important one to be, viz. that all the choir of heaven and furniture of the earth, in a word all those bodies which compose the mighty frame of the world, have not any subsistence without a mind, that their being is to be perceiv'd or known; that consequently so long as they are not actually perceiv'd by me, or do not exist in my mind or that of any other created spirit, they must either have no existence at all, or else subsist in the mind of some eternal spirit: it being perfectly unintelligible and involving all the absurdity of

abstraction, to attribute to any single part of them an existence independent of a spirit.

To be convinced of which, the reader need only reflect and try to separate in his own thoughts the being of a sensible thing from its being perceived.

7. From what has been said, it follows there is not any other substance than spirit or that which perceives. But for the fuller proof of this point, let it be consider'd, the sensible qualities are colour, figure, motion, smell, taste, and such like, that is, the ideas perceiv'd by sense. Now for an idea to exist in an unperceiving thing is a manifest contradiction, for to have an idea is all one as to perceive, that therefore wherein colour, figure, and the like qualities exist must perceive them; hence 'tis clear there can be no unthinking substance or *substratum* of those ideas.

8. But say you, thô the ideas themselves do not exist without the mind, yet there may be things like them whereof they are copies or resemblances, which things exist without the mind, in an unthinking substance. I answer an idea can be like nothing but an idea, a colour, or figure, can be like nothing but another colour or figure. If we look but ever so little into our thoughts, we shall find it impossible for us to conceive a likeness except only between our ideas. Again, I ask whether those suppos'd originals or external things, of which our ideas are the pictures or representations, be themselves perceivable or no? If they are, then they are ideas and we have gain'd our point; but if you say they are not, I appeal to any one whether it be sense, to assert a colour is like something which is invisible; hard or soft, like something which is intangible, and so of the rest.

35. From DAVID HUME'S *An Enquiry concerning Human Understanding* (1748)

[Hume maintains that the Law of Cause and Effect has no logical basis and that all we can know are our own ideas.]

When any natural object or event is presented, it is impossible for us, by any sagacity or penetration, to discover, or even conjecture, without experience, what event will result from it, or to carry our foresight beyond that object which is immediately present to the memory and senses. Even after one instance or experiment where we have observed a particular event to follow upon another, we are not entitled to form a general rule, or foretell what will happen in like cases; it being justly esteemed an unpardonable temerity to judge of the whole course of nature from one single experiment, however accurate or certain. But when one particular species of event has always, in all instances, been conjoined with another, we make no longer any scruple of foretelling one upon the appearance of the other, and of employing that reasoning, which can alone assure us of any matter of fact or existence. We then call the one object, *Cause*; the other, *Effect*. We suppose that there is some connexion between them; some power in the one, by which it infallibly produces the other, and operates with the greatest certainty and strongest necessity.

It appears, then, that this idea of a necessary connexion among events arises from a number of similar instances which occur of the constant conjunction of these events; nor can that idea ever be suggested by any one of these instances, surveyed in all possible lights and positions. But there is nothing in a number of instances different from every single instance, which is supposed to be exactly similar; except only, that after a repetition of similar instances, the mind is carried by habit, upon the appearance of one event, to expect its usual attendant, and to believe that it will exist. This connexion, therefore, which we *feel* in the mind, this customary transition of the imagination from one object to its usual attendant, is the sentiment or impression from which we form the idea of power or necessary connexion. Nothing farther is in the case. Contemplate the subject on all sides; you will never find any other origin of that idea. This is the sole difference between one instance, from which we can never receive the idea of connexion, and a number of similar instances, by which it is suggested. The first time a man saw the communication of motion by impulse, as by the shock of two billiard balls, he could not pronounce that the one event was *connected*: but only that it was *conjoined* with the other. After he has observed several instances of this nature, he then pronounces them to be *connected*. What alteration has happened to give rise to this new idea of *connexion*? Nothing but that he now *feels* these events to be *connected* in his imagination, and can readily foretell the existence of one from the appearance of the other. When we say, therefore, that one object is connected with another, we mean only that they have acquired a connexion in our thought, and give rise to this inference, by which they become proofs of each other's existence: A conclusion which is somewhat extraordinary, but which seems founded on sufficient evidence. . . .

By what argument can it be proved that the perceptions of the mind must be caused by external objects entirely different from them, though resembling them (if that be possible) and could not arise either from the energy of the mind itself, or from the suggestion of some invisible and unknown spirit, or from some other cause still more unknown to us? It is acknowledged that, in fact, many of these perceptions arise not from anything external, as in dreams, madness, and other diseases. And nothing can be more inexplicable than the manner in which body should so operate upon mind as ever to convey an image of itself to a substance supposed of so different and even contrary a nature.

It is a question of fact, whether the perceptions of the senses be produced by external objects resembling them: how shall this question be determined? By experience, surely; as all other questions of a like nature. But here experience is, and must be, entirely silent. The mind has never anything present to it but the perceptions and cannot possibly reach any experience of their connexion with objects. The supposition of such a connexion is, therefore, without any foundation in reasoning.

36. From IMMANUEL KANT's *Critique of Pure Reason* (1787)

[Kant discusses the limitations of empiricism and introduces his views on *a priori* concepts or categories.]

The illustrious Locke, . . . meeting with pure concepts of the understanding in experience, deduced them also from experience, and yet proceeded so *inconsequently* that he attempted with their aid to obtain knowledge which far transcends all limits of experience. David Hume recognised that, in order to be able to do this, it was necessary that these concepts should have an *a priori* origin. But since he could not explain how it can be possible that the understanding must think concepts, which are not in themselves connected in the understanding, as being necessarily connected in

the object, and since it never occurred to him that the understanding might itself, perhaps, through these concepts, be the
author of the experience in which its objects are found, he was
constrained to derive them from experience, namely, from a
subjective necessity (that is, from *custom*), which arises from
repeated association in experience, and which comes mistakenly
to be regarded as objective. But from these premises he argued
quite consistently. It is impossible, he declared, with these
concepts and the principles to which they give rise, to pass beyond
the limits of experience. Now this *empirical* derivation, in which
both philosophers agree, cannot be reconciled with the scientific
a priori knowledge which we do actually possess, namely, *pure
mathematics* and *general science of nature*; and this fact therefore
suffices to disprove such derivation.

While the former of these two illustrious men opened a wide
door to *enthusiasm*—for if reason once be allowed such rights, it
will no longer allow itself to be kept within bounds by vaguely
defined recommendations of moderation—the other gave himself
over entirely to *scepticism*, having, as he believed, discovered that
what had hitherto been regarded as reason was but an all-
prevalent illusion infecting our faculty of knowledge. We now
propose to make trial whether it be not possible to find for human
reason safe conduct between these two rocks, assigning to her
determinate limits, and yet keeping open for her the whole field
of her appropriate activities. . . .

If the objects with which our knowledge has to deal were things
in themselves, we could have no *a priori* concepts of them. For
from what source could we obtain the concepts? If we derived
them from the object (leaving aside the question how the object
could become known to us), our concepts would be merely
empirical, not *a priori*. And if we derived them from the self, that
which is merely in us could not determine the character of an
object distinct from our representations, that is, could not be
a ground why a thing should exist characterised by that which we
have in our thought, and why such a representation should not,
rather, be altogether empty. But if, on the other hand, we have
to deal only with appearances, it is not merely possible, but
necessary, that certain *a priori* concepts should precede empirical
knowledge of objects. For since a mere modification of our
sensibility can never be met with outside us, the objects, as

appearances, constitute an object which is merely in us. Now to assert in this manner, that all these appearances, and consequently all objects with which we can occupy ourselves, are one and all in me, that is, are determinations of my identical self, is only another way of saying that there must be a complete unity of them in one and the same apperception. But this unity of possible consciousness also constitutes the form of all knowledge of objects; through it the manifold is thought as belonging to a single object. Thus the mode in which the manifold of sensible representation (intuition) belongs to one consciousness precedes all knowledge of the object as the intellectual form of such knowledge, and itself constitutes a formal *a priori* knowledge of all objects, so far as they are thought (categories). The synthesis of the manifold through pure imagination, the unity of all representations in relation to original apperception, precede all empirical knowledge. Pure concepts of understanding are thus *a priori* possible, and, in relation to experience, are indeed necessary; and this for the reason that our knowledge has to deal solely with appearances, the possibility of which lies in ourselves, and the connection and unity of which (in the representation of an object) are to be met with only in ourselves. Such connection and unity must therefore precede all experience, and are required for the very possibility of it in its formal aspect. From this point of view, the only feasible one, our deduction of the categories has been developed.

37. From DENIS DIDEROT'S *Conversation of a Philosopher with the Maréchale de X* (1776)

[In this amusing dialogue Crudeli, who represents Diderot's views, suggests that mind is a product of matter, and that man is a machine.]

I had some business or other with the Maréchal de X. I went to his mansion one morning; he was absent; I had myself announced to Madame la Maréchale. She is a charming woman; she is as beautiful and as devout as an angel; sweetness is clearly expressed on her countenance; and she has, moreover, a tone of voice and a candour in discussion quite in keeping with her expression. She was at her toilette. A chair is drawn up for me; I seat myself and we chat....

The Maréchale: Aren't you Monsieur Crudeli?

Crudeli. Yes, Madame.

The M. Then it's you who believes in nothing?

Cr. The same....

The M. But this world of ours, who made it?

Cr. I ask you that.

The M. God made it.

Cr. And what is God?

The M. A spirit.

Cr. If a spirit can make matter, why could not matter make a spirit?

The M. But why should it make it?

Cr. I see it do it every day. Do you believe that animals have souls?

The M. Certainly I believe it.

Cr. And would you tell me what becomes of the soul of a Peruvian serpent, for example, while it becomes dried, hung in a chimney, exposed to the smoke for a year or two continuously?

The M. I don't care what happens to it; what has it to do with me?

Cr. Madame la Maréchale doesn't know that this dried and smoked serpent revives and is alive again.

The M. I believe none of that.

Cr. It is a clever man, however, Bouguer, who asserts it.

The M. Your clever man has lied about it.

Cr. And if he had spoken truly?

The M. I should have to believe that animals are machines.

Cr. And man, who is only an animal a little more perfect than others....But M. le Maréchal...

The M. Just one more question and it's the last. Are you really quite at peace in your unbelief?

Cr. One could not be more so.

The M. But what if you've deceived yourself?

Cr. If I should have deceived myself?

The M. All that you believed false would be true, and you would be damned. Monsieur Crudeli, it is a dreadful thing to be damned; to burn for all eternity, that's terribly long.

38. From VOLTAIRE'S *A Treatise on Toleration* (1763)

[One of the main ideals of the Enlightenment was toleration. The wars of religion were over but men still tortured and executed each other because of religious prejudice and hysteria. Today men are murdered, on a much larger scale, for ideological and racial reasons. In the following characteristic *conte* Voltaire mocks the absurdity of the conflicts between Christian sects.]

In the early years of the reign of the great Emperor Kam-hi a mandarin of the city of Canton heard from his house a great noise, which proceeded from the next house. He inquired if anybody was being killed, and was told that the almoner of the Danish missionary society, a chaplain from Batavia, and a Jesuit were disputing. He had them brought to his house, put tea and sweets before them, and asked why they quarrelled.

The Jesuit replied that it was very painful for him, since he was always right, to have to do with men who were always wrong; that he had at first argued with the greatest restraint, but had at length lost patience.

The mandarin, with the utmost discretion, reminded them that politeness was needed in all discussion, told them that in China men never became angry, and asked the cause of the dispute.

The Jesuit answered: 'My lord, I leave it to you to decide. These two gentlemen refuse to submit to the decrees of the Council of Trent.'

'I am astonished', said the mandarin. Then, turning to the refractory pair, he said: 'Gentlemen, you ought to respect the opinions of a large gathering. I do not know what the Council of Trent is, but a number of men are always better informed than a single one. No one ought to imagine that he is better than others, and has a monopoly of reason. So our great Confucius teaches; and, believe me, you will do well to submit to the Council of Trent.'

The Dane then spoke. 'My lord speaks with the greatest wisdom', he said; 'we respect great councils, as is proper, and therefore we are in entire agreement with several that were held before the Council of Trent.'

'Oh, if that is the case', said the mandarin, 'I beg your pardon. You may be right. So you and this Dutchman are of the same opinion, against this poor Jesuit.'

'Not a bit', said the Dutchman. 'This fellow's opinions are

almost as extravagant as those of the Jesuit yonder, who has been so very amiable to you. I can't bear them.'

'I don't understand', said the mandarin. 'Are you not all three Christians? Have you not all three come to teach Christianity in our empire? Ought you not, therefore, to hold the same dogmas?'

'It is this way, my lord', said the Jesuit; 'these two are mortal enemies, and are both against me. Hence it is clear that they are both wrong, and I am right.'

'That is not quite clear', said the mandarin; 'strictly speaking, all three of you may be wrong. I should like to hear you all, one after the other.'

The Jesuit then made a rather long speech, during which the Dane and the Dutchman shrugged their shoulders. The mandarin did not understand a word of it. Then the Dane spoke; the two opponents regarded him with pity, and the mandarin again failed to understand. The Dutchman had the same effect. In the end they all spoke together and abused each other roundly. The good mandarin secured silence with great difficulty, and said: 'If you want us to tolerate your teaching here, begin by being yourselves neither intolerant nor intolerable.'

When they went out the Jesuit met a Dominican friar, and told him that he had won, adding that truth always triumphed. The Dominican said: 'Had I been there, you would not have won; I should have convicted you of lying and idolatry.' The quarrel became warm, and the Jesuit and Dominican took to pulling each other's hair. The mandarin, on hearing of the scandal, sent them both to prison. A sub-mandarin said to the judge: 'How long does your excellency wish them to be kept in prison?' 'Until they agree', said the judge. 'Then', said the sub-mandarin, 'they are in prison for life.' 'In that case,' said the judge, 'until they forgive each other.' 'They will never forgive each other', said the other; 'I know them.' 'Then', said the mandarin, 'let them stop there until they pretend to forgive each other.'

39. From *Conversations of Goethe with Eckermann and Soret* (1823)

[Goethe talks about his theory of colours—now known to be scientifically worthless.]

Tuesday, 30 December 1823

We then talked about the natural sciences, especially about the narrow-mindedness with which learned men contend amongst themselves for priority. 'There is nothing,' said Goethe, 'through which I have learned to know mankind better, than through my philosophical exertions. It has cost me a great deal, and has been attended with great annoyance, but I nevertheless rejoice that I have gained the experience....'

'A Frenchman said to a friend of mine, concerning my theory of colours,—"We have worked for fifty years to establish and strengthen the kingdom of Newton, and it will require fifty years more to overthrow it." The body of mathematicians has endeavoured to make my name so suspected in science that people are afraid of even mentioning it. Some time ago, a pamphlet fell into my hands, in which subjects connected with the theory of colours were treated: the author appeared quite imbued with my theory, and had deduced everything from the same fundamental principles. I read the publication with great delight, but, to my no small surprise, found that the author did not once mention my name. The enigma was afterwards solved. A mutual friend called on me, and confessed to me that the clever young author had wished to establish his reputation by the pamphlet, and had justly feared to compromise himself with the learned world, if he ventured to support by my name the views he was expounding. The little pamphlet was successful, and the ingenious young author has since introduced himself to me personally, and made his excuses.'

'This circumstance appears to me the more remarkable,' said I, 'because in everything else people have reason to be proud of you as an authority, and every one esteems himself fortunate who has the powerful protection of your public countenance. With respect to your theory of colours, the misfortune appears to be, that you have to deal not only with the renowned and universally acknowledged Newton, but also with his disciples, who are spread all over the world, who adhere to their master, and whose name is legion. Even supposing that you carry your point at last, you will certainly for a long space of time stand alone with your new theory.'

'I am accustomed to it, and prepared for it', returned Goethe. 'But say yourself,' continued he, 'have I not had sufficient reason to feel proud, when for twenty years I have been forced to own to myself that the great Newton, and all mathematicians and august calculators with him, have fallen into a decided error respecting the theory of colours; and that I, amongst millions, am the only one who knows the truth on this important subject? With this feeling of superiority, it was possible for me to bear with the stupid pretensions of my opponents. People endeavoured to attack me and my theory in every way, and to render my ideas ridiculous; but, nevertheless, I rejoiced exceedingly over my completed work. All the attacks of my adversaries only serve to expose to me the weakness of mankind.'

While Goethe spoke thus, with such a force and a fluency of expression as I have not the power to reproduce with perfect truth, his eyes sparkled with unusual fire; an expression of triumph was observable in them; whilst an ironical smile played upon his lips. The features of his fine countenance were more imposing than ever.

CHAPTER II

THE ATOMIC THEORY

40. From LUCRETIUS' *De Rerum Natura* (On the Nature of Things)

[Lucretius expounded an atomic theory. In the first part of this passage the atoms are referred to as germs.]

> ...no rest is ever found
> For germs throughout the void, but driven on
> In ceaseless varied motion some rebound,
> Leaving large gaps, while some are knit together
> With hardly any interspace at all:
> And these more closely bound with little space
> Locked close by their own intertangled forms,
> These form the rocks, the unyielding iron mass,
> And things like these: but those which spring apart
> Rebounding with great intervals between,
> These give us the thin air...
>   for from our senses far
> The nature of these primal atoms lies.
> Since they're beyond our sight, their motions too
> Must be beyond our ken, and all the more
> Since what you see its movement oft conceals,
> By the very distance from us that it lies.
> Thus oft in the hillside the woolly flocks,
> Cropping the gladsome mead, creep slowly in,
> Where'er the grass with pearly dew invites,
> And lambs full-fed sport round, and butt each other
> In sparkling play: you only see the mass,
> It rests on the green hill a spot of white.

41. From JOHN DALTON'S *A New System of Chemical Philo-sophy* (1808)

[Dalton introduces the first table of atomic and molecular weights.]

In all chemical investigations, it has justly been considered an important object to ascertain the relative *weights* of the simples which constitute a compound. But unfortunately the enquiry has terminated here; whereas, from the relative weights in the mass, the relative weights of the ultimate particles, or atoms of the bodies might have been inferred, from which their number and weight in various other compounds would appear, in order to assist and guide future investigations, and to correct their results. Now it is one great object of this work, to show the importance and advantage of ascertaining *the relative weights of the ultimate particles, both of simple and compound bodies, the number of simple elementary particles which constitute one compound particle, and the number of less compound particles which enter into the formation of one more compound particle.*

[Dalton then proceeds to give the rules by which he decides how many atoms are contained in a compound particle. These may be termed rules of maximum simplicity. In the case of a substance like water, he assumes that one atom of hydrogen combines with one atom of oxygen. In the cases of the two compounds formed by carbon and oxygen, namely carbonic oxide (carbon monoxide) and carbonic acid (carbon dioxide), he assumes that the former is composed of one atom of carbon and one atom of oxygen, while the latter is composed of one atom of carbon and two atoms of oxygen. And so on.]

From the novelty as well as importance of the ideas suggested in this chapter, it is deemed expedient to give plates, exhibiting the mode of combination in some of the more simple cases. A specimen of these accompanies this first part. The elements or atoms of such bodies as are conceived at present to be simple, are denoted by a small circle, and some distinctive mark; and the combinations consist in the juxtaposition of two or more of these; when three or more particles of elastic fluids are combined together in one, it is to be supposed that the particles of the same kind repel each other, and therefore take their stations accordingly.

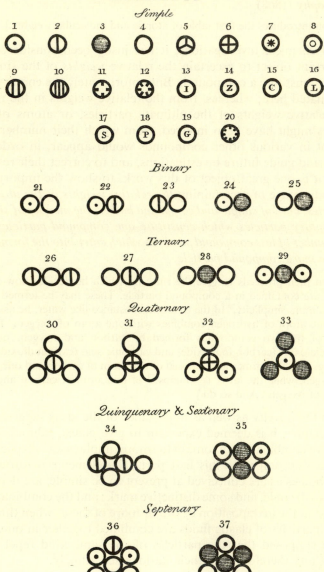

ELEMENTS

Simple

Binary

Ternary

Quaternary

Quinquenary & Sextenary

Septenary

Plate IV. This plate contains the arbitrary marks or signs chosen to represent the several chemical elements or ultimate particles.

Fig.			Fig.		
1	Hydrog. its rel. weight	1	11	Strontites	46
2	Azote	5	12	Barytes	68
3	Carbone or charcoal	5	13	Iron..............	38
4	Oxygen	7	14	Zinc	56
5	Phosphorus	9	15	Copper	56
6	Sulphur	13	16	Lead	95
7	Magnesia	20	17	Silver	100
8	Lime	23	18	Platina	100
9	Soda	28	19	Gold	140
10	Potash	42	20	Mercury	167

21 An atom of water or steam, composed of 1 of oxygen and 1 of hydrogen, retained in physical contact by a strong affinity, and supposed to be surrounded by a common atmosphere of heat; its relative weight 8

22 An atom of ammonia, composed of 1 of azote and 1 of hydrogen 6

23 An atom of nitrous gas, composed of 1 of azote and 1 of oxygen 12

24 An atom of olefiant gas, composed of 1 of carbone and 1 of hydrogen 6

25 An atom of carbonic oxide composed of 1 of carbone and 1 oxygen... 12

26 An atom of nitrous oxide, 2 azote + 1 oxygen 17

27 An atom of nitric acid, 1 azote + 2 oxygen 19

28 An atom of carbonic acid, 1 carbone + 2 oxygen....... 19

29 An atom of carburetted hydrogen, 1 carbone + 2 hydrogen 7

30 An atom of oxynitric acid, 1 azote + 3 oxygen 26

31 An atom of sulphuric acid, 1 sulphur + 3 oxygen 34

32 An atom of sulphuretted hydrogen, 1 sulphur + 3 hydrogen 16

33 An atom of alcohol, 3 carbone + 1 hydrogen 16

34 An atom of nitrous acid, 1 nitric acid + 1 nitrous gas .. 31

35 An atom of acetous acid, 2 carbone + 2 water 26

36 An atom of nitrate of ammonia, 1 nitric acid + 1 ammonia + 1 water 33

37 An atom of sugar, 1 alcohol + 1 carbonic acid 35

42. From AMADEO AVOGADRO'S *Essay on a Manner of Determining the Relative Masses of the Elementary Molecules of Bodies, and the Proportions in which they enter into these Compounds* (1811)

[Avogadro's hypothesis that equal volumes of all gases, at the same temperature and pressure, contain the same number of molecules, is one of the chief foundation stones on which chemistry is built. A molecule of hydrogen is H_2 whereas an atom is H. Avogadro is quite clear about this distinction although his nomenclature is not ideal. He uses the term *molecule* for either the modern atom or molecule; *simple molecule* or *elementary molecule* for atom; *composite molecule* for molecule; and *integral molecule* for the molecule of a compound.

In the last part of this extract Avogadro points out that Dalton assumed water to be HO instead of H_2O and hence obtained an atomic weight for oxygen only half its true value.]

M. Gay-Lussac has shown in an interesting Memoir (Mémoires de la Société d'Arcueil, Tome II.) that gases always unite in a very simple proportion by volume, and that when the result of the union is a gas, its volume also is very simply related to those of its components. But the quantitative proportions of substances in compounds seem only to depend on the relative number of molecules which combine, and on the number of composite molecules which result. It must then be admitted that very simple relations also exist between the volumes of gaseous substances and the numbers of simple or compound molecules which form them. The first hypothesis to present itself in this connection, and apparently even the only admissible one, is the supposition that the number of integral molecules in any gases is always the same for equal volumes, or always proportional to the volumes. Indeed, if we were to suppose that the number of molecules contained in a given volume were different for different gases, it would scarcely be possible to conceive that the law regulating the distance of molecules could give in all cases relations so simple as those which the facts just detailed compel us to acknowledge between the volume and the number of molecules....

Setting out from this hypothesis, it is apparent that we have the means of determining very easily the relative masses of the molecules of substances obtainable in the gaseous state, and the relative number of these molecules in compounds; for the ratios of the masses of molecules are then the same as those of the densities of the different gases at equal temperature and pressure, and the relative number of molecules in a compound is given at once by the ratio of the volumes of the gases that form it. For example, since the numbers 1·10359 and 0·07321 express the densities of the two gases oxygen and hydrogen compared to that of atmospheric air as unity, and the ratio of the two numbers consequently represents the ratio between the masses of equal volumes of these two gases, it will also represent on our hypothesis the ratio of the masses of their molecules. Thus the mass of the molecule of oxygen will be about 15 times that of the molecule of hydrogen, or more exactly, as 15·074 to 1. . . .

Dalton, on arbitrary suppositions as to the most likely relative number of molecules in compounds, has endeavoured to fix ratios between the masses of the molecules of simple substances. Our hypothesis, supposing it well-founded, puts us in a position to confirm or rectify his results from precise data, and, above all, to assign the magnitude of compound molecules according to the volumes of the gaseous compounds, which depend partly on the division of molecules entirely unsuspected by this physicist.

Thus Dalton supposes that water is formed by the union of hydrogen and oxygen, molecule to molecule. From this, and from the ratio by weight of the two components, it would follow that the mass of the molecule of oxygen would be to that of hydrogen as $7\frac{1}{2}$ to 1 nearly, or, according to Dalton's evaluation, as 6 to 1. This ratio on our hypothesis is, as we saw, twice as great, namely, as 15 to 1. As for the molecule of water, its mass ought to be roughly expressed by $15 + 2 = 17$ (taking for unity that of hydrogen), if there were no division of the molecule into two; but on account of this division it is reduced to half, $8\frac{1}{2}$, or more exactly 8·537, as may also be found directly by dividing the density of aqueous vapour 0·625 (Gay-Lussac) by the density of hydrogen 0·0732. This mass only differs from 7, that assigned to it by Dalton, by the difference in the values for the composition of water; so that in this respect Dalton's result is approximately correct from the combination of two compensating errors,—the error in the mass of the molecule of oxygen, and his neglect of the division of the molecule.

CHAPTER 12

THE WAVE THEORY OF LIGHT

43. From CHRISTIAAN HUYGENS' *Treatise on Light* (1690)

[Huygens argues that light, like sound, is a form of wave motion.]

It is inconceivable to doubt that light consists in the motion of some sort of matter. For whether one considers its production, one sees that here upon the Earth it is chiefly engendered by fire and flame which contain without doubt bodies that are in rapid motion, since they dissolve and melt many other bodies, even the most solid; or whether one considers its effects, one sees that when light is collected, as by concave mirrors, it has the property of burning as a fire does, that is to say it disunites the particles of bodies. This is assuredly the mark of motion, at least in the true Philosophy, in which one conceives the causes of all natural effects in terms of mechanical motions. This, in my opinion, we must necessarily do, or else renounce all hopes of ever comprehending anything in Physics.

And as, according to this Philosophy, one holds as certain that the sensation of sight is excited only by the impression of some movement of a kind of matter which acts on the nerves at the back of our eyes, there is here yet one reason more for believing that light consists in a movement of the matter which exists between us and the luminous body.

Further, when one considers the extreme speed with which light spreads on every side, and how, when it comes from different regions, even from those directly opposite, the rays traverse one another without hindrance, one may well understand that when we see a luminous object, it cannot be by any transport of matter coming to us from this object, in the way in which a shot or an arrow traverses the air; for assuredly that would too greatly impugn these two properties of light, especially the second of them. It is then in some other way that light spreads; and that which can lead us to comprehend it is the knowledge which we have of the spreading of Sound in the air.

We know that by means of the air, which is an invisible and impalpable body, Sound spreads around the spot where it has been produced, by a movement which is passed on successively from one part of the air to another; and that the spreading of this movement, taking place equally rapidly on all sides, ought to form spherical surfaces ever enlarging and which strike our ears. Now there is no doubt at all that light also comes from the luminous body to our eyes by some movement impressed on the matter which is between the two; since, as we have already seen, it cannot be by the transport of a body which passes from one to the other. If, in addition, light takes time for its passage—which we are now going to examine—it will follow that this movement, impressed on the intervening matter, is successive; and consequently it spreads, as Sound does, by spherical surfaces and waves: for I call them waves from their resemblance to those which are seen to be formed in water when a stone is thrown into it, and which present a successive spreading as circles, though these arise from another cause, and are only in a flat surface.

44. From ISAAC NEWTON'S *Opticks* (1704)

[This is part of Query 28, in which Newton argues that light cannot consist of waves because waves spread round corners, whereas light travels in straight lines.]

Query 28.   Are not all Hypotheses erroneous, in which Light is supposed to consist in Pression or Motion, propagated through a fluid Medium?....For Pression or Motion, cannot be propagated in a Fluid in right Lines, beyond an Obstacle which stops part of the Motion, but will bend and spread every way into the quiescent Medium which lies beyond the Obstacle....The Waves on the Surface of stagnating Water, passing by the sides of a broad Obstacle which stops part of them, bend afterwards and dilate themselves gradually into the quiet Water behind the Obstacle. The Waves, Pulses or Vibrations of the Air, wherein Sounds consist, bend manifestly, though not so much as the Waves of Water. For a Bell or a Cannon may be heard beyond a Hill which intercepts the sight of the sounding Body, and Sounds are propagated as readily through crooked Pipes as through streight ones. But Light is never known to follow crooked Passages nor to bend into the Shadow.

[In Query 29 Newton suggests a corpuscular theory of light.]

Query 29.   Are not the Rays of Light very small Bodies emitted from shining Substances? For such Bodies will pass through uniform Mediums in right Lines without bending into the Shadow, which is the Nature of the Rays of Light. They will also be capable of several Properties, and be able to conserve their Properties unchanged in passing through several Mediums, which is another Condition of the Rays of Light.

[In Query 17 Newton supplements the corpuscular theory with a wave theory.]

Query 17.   If a stone be thrown into stagnating Water, the Waves excited thereby continue some time to arise in the place where the Stone fell into the Water, and are propagated from thence in concentrick Circles upon the Surface of the Water to great distances. And the Vibrations or Tremors excited in the Air by percussion, continue a little time to move from the place of percussion in concentrick Spheres to great distances. And in like manner, when a Ray of Light falls upon the Surface of any pellucid Body, and is there refracted or reflected, may not Waves of Vibrations, or Tremors, be thereby excited in the refracting or reflecting Medium at the point of Incidence, and continue to arise there, and to be propagated from thence as long as they continue to arise and be propagated, when they are excited in the bottom of the Eye by the Pressure or Motion of the Finger, or by the Light which comes from the Coal of Fire in the Experiments above mention'd? and are not these Vibrations propagated from the point of Incidence to great distances? And do they not over-take the Rays of Light, and by overtaking them successively, do they not put them into the Fits of easy Reflexion and easy Trans-mission described above? For if the Rays endeavour to recede from the densest part of the Vibration, they may be alternately accelerated and retarded by the Vibrations overtaking them.

[In Queries 12, 13 and 14 Newton suggests that sensation of colour is determined by wavelength.]

Query 12.   Do not the Rays of Light in falling upon the bottom of the Eye excite Vibrations in the *Tunica Retina*? Which Vibra-tions, being propagated along the solid Fibres of the optick Nerves into the Brain, cause the Sense of seeing....

Query 13. Do not several sorts of Rays make Vibrations of several bignesses, which according to their bignesses excite Sensations of several Colours, much after the manner that the Vibrations of the Air, according to their several bignesses excite Sensations of several Sounds? And particularly do not the most refrangible Rays excite the shortest Vibrations for making a Sensation of deep violet, the least refrangible the largest for making a Sensation of deep red, and the several intermediate sorts of Rays, Vibrations of several intermediate bignesses to make Sensations of the several intermediate Colours?

Query 14. May not the harmony and discord of Colours arise from the proportions of the Vibrations propagated through the Fibres of the optick Nerves into the Brain, as the harmony and discord of Sounds arise from the proportions of the Vibrations of the Air?

45. From Dr THOMAS YOUNG'S *Reply to the animadversions of the Edinburgh Reviewers on some papers published in the Philosophical Transactions* (1804)

[This is the opening of the rather prolix and ponderous reply of Young to the venomous attacks made upon his work by Henry Brougham. It is notable for the statement of the principle of interference at the end of the passsage.]

Welbeck Street, 30 Nov. 1804

A man who has a proper regard for the dignity of his own character, although his sensibility may sometimes be awakened by the unjust attacks of interested malevolence, will esteem it in general more advisable to bear, in silence, the temporary effects of a short-lived injury, than to suffer his own pursuits to be interrupted, in making an effort to repel the invective, and to punish the aggressor. But it is possible that art and malice may be so insidiously combined, as to give to the grossest misrepresentations the semblance of justice and candour; and, especially where the subject of the discussion is of a nature little adapted to the comprehension of the generality of readers, even a man's friends may be so far misled by a garbled extract from his own works, and by the specious mixture of partial truth with essential falsehood, that they may not only be unable to defend him from the unfavourable opinion of others, but may themselves be disposed to suspect, in spite of their partiality, that he has been hasty and inconsiderate at least, if not radically weak and mistaken. In such a case, he owes to his friends such explanations as will enable them to see clearly the injustice of the accusation, and the iniquity of its author: and, if he is in a situation which requires that he should in a certain degree possess the public confidence, he owes to himself and to the public to prove, that the charges of imbecility of mind and perversity of disposition are not more founded with regard to him, than with regard to all who are partakers with him in the unavoidable imperfections of human nature.

Precisely such is my situation. I have at various times communicated to the Royal Society, in a very abridged form, the results of my experiments and investigations, relating to different branches of natural philosophy: and the Council of the Society, with a view perhaps of encouraging patient diligence, has honoured my essays with a place in their Transactions. Several of these essays have been singled out, in an unprecedented manner, from the volumes in which they were printed, and have been made the subjects, in the second and ninth numbers of the Edinburgh Review, not of criticism, but of ridicule and invective; of an attack, not only upon my writings and my literary pursuits, but almost on my moral character. The peculiarity of the style and tendency of this attack led me at once to suspect, that it must have been suggested by some other motive than the love of truth; and I have both internal and external evidence for believing, that the articles in question are, either wholly, or in great measure, the productions of an individual, upon whose mathematical works I had formerly thought it necessary to make some remarks, which, though not favourable, were far from being severe....

I have indeed been accused of insinuating 'that Sir Isaac Newton was but a sorry philosopher'. But it is impossible that an impartial person should read my essays on the subject of light without being sensible that I have as high a respect for his unparalleled talents and acquirements as the blindest of his followers, and the most parasitical of his defenders. I have acknowledged that 'his merits are great beyond all contest or comparison'; that 'his discovery of the composition of white light would alone have immortalised his name'; that the very arguments which tend to overthrow his hypothesis respecting the emanation of light, 'give the strongest proofs of the admirable accuracy of his experiments'; and that a person may, 'with the greatest justice, be attached to every doctrine which is stamped with the Newtonian approbation'. The printer of the Review, feeling perhaps that the last expressions would militate too much in my favour, has thought fit to plunder me of them, by omitting the marks of quotation and to attribute them to my antagonist. But, much as I venerate the name of Newton, I am not therefore obliged to believe that he was infallible. I see, not with exultation, but with regret, that he was liable to err, and that his authority has, perhaps, sometimes even retarded the progress of Science. . . .

It was in May 1801 that I discovered, by reflecting on the beautiful experiments of Newton, a law which appears to me to account for a greater variety of interesting phenomena than any other optical principle that has yet been made known. I shall endeavour to explain this law by a comparison.

Suppose a number of equal waves of water to move upon the surface of a stagnant lake, with a certain constant velocity, and to enter a narrow channel leading out of the lake. Suppose then another similar cause to have excited another equal series of waves, which arrive at the same channel, with the same velocity, and at the same time with the first. Neither series of waves will destroy the other, but their effects will be combined: if they enter the channel in such a manner that the elevations of one series coincide with those of the other, they must together produce a series of greater joint elevations; but if the elevations of one series are so situated as to correspond to the depressions of the other, they must exactly fill up those depressions, and the surface of the water must remain smooth; at least I can discover no alternative, either from theory or from experiment.

Now I maintain that similar effects take place whenever two portions of light are thus mixed; and this I call the general law of the interference of light. I have shown that this law agrees, most accurately, with the measures recorded in Newton's *Optics*, relative to the colours of transparent substances, observed under circumstances which had never before been subjected to calculation, and with a great diversity of other experiments never before explained. This, I assert, is a most powerful argument in favour of the theory which I had before revived.

46. From AUGUSTIN FRESNEL'S *Memoir on the Diffraction of Light* (1819)

[In this opening passage of the memoir Fresnel points out that the wave theory of light is far more economical and convincing in its basic assumptions than the corpuscular or emission theory.]

Before I concern myself particularly with the numerous and varied phenomena included under the common name of diffraction I feel that I must put forward several general considerations on the two systems which, until now, have divided men of science on the nature of light. Newton assumed that particles of light, shot from luminous bodies, come straight to our eyes, where they produce by their onset the sensation of sight. Descartes, Hooke, Huyghens and Euler thought that light resulted from the vibrations of an universal, extremely subtle fluid, agitated by the rapid movements of the particles of the luminous body, in the same way that the air is disturbed by the vibrations of sounding bodies; so that, in this system, it is not the molecules of the fluid in contact with the luminous body which reach the organ of sight, but only the movement which has been impressed upon them.

The first hypothesis has the advantage of leading to more evident consequences because the mathematical analysis is applied more easily; the second, on the contrary, presents in this respect great difficulties. But, in the choice of a system, one must have regard only to the simplicity of the hypotheses; that of the calculus must not have any weight in the balance of probabilities. Nature is not troubled by difficulties of analysis; she shuns only complication of means. She appears to be purposed to make much with little; this is a principle of which new proofs are unceasingly forthcoming from the development of the physical sciences. Astronomy, the glory of the human mind, above all provides a striking confirmation; all Kepler's laws were traced back by the genius of Newton to the single law of gravitation, which then served to explain and even to reveal the most complicated and least apparent perturbations of the planetary movements.

If men have sometimes gone astray in wishing to simplify the elements of a science, it is because they have established systems before collecting a sufficiently large number of facts. An hypothesis of the kind they put forward, very simple when only one class of phenomena is considered, necessitates many other hypotheses when they wish to break out of the tight circle in which they have been confined. If nature has purposed to produce the *maximum* of effects with the *minimum* of causes, it is in the harmony of her laws that she has been able to resolve this great problem.

It is undoubtedly difficult to discover the basis of this admirable economy, namely the simplest causes of the phenomena as seen from an extended viewpoint. But if this general principle of the philosophy of the physical sciences does not lead immediately to the knowledge of truth, it can nevertheless direct the efforts of the human mind, in the rejection of systems which account for the phenomena by too large a number of different causes, and in the adoption by preference of those which, resting on the least number of hypotheses, are consequently the most fertile.

In this respect, the system which makes light consist of the vibrations of an universal fluid has great advantages over that of emission. It enables us to understand how light is susceptible of undergoing so many diverse modifications. I do not mean here those which it experiences momentarily in the bodies which it traverses and which can always be explained by the nature of these media, but I wish to speak of those permanent modifications which it carries with it and which stamp it with new characteristics. One conceives that a fluid, an assemblage of an infinity of mobile molecules subject to a mutual dependence, is capable of a great number of different modifications, by reason of the relative movements which are impressed upon them. The vibrations of the air and the variety of the sensations which they produce on the organ of hearing offer a remarkable parallel.

In the emission system, on the other hand, the motion of each light particle being independent of that of the others, the number of modifications to which it is susceptible appears extremely limited. One can add a movement of rotation to that of transmission; but that is all. As for oscillatory movements, their existence is conceivable only in media which maintain an unequal action on the different sides of the light particles, supposed endowed with different properties. As soon as this action ceases, the oscillations must also cease or become transformed into rotary movements. Thus the rotary movements and the diversity of the sides of a light particle are the only mechanical resources of the emission theory to represent all the permanent modifications of light. They appear very inadequate, if one pays attention to the multitude of phenomena which optics offers. One is more convinced of this on reading the Treatise on experimental and mathematical physics of M. Biot, in which are developed, with great detail and clarity, the principal consequences of Newton's system. One will find there, to account for the phenomena, that it is necessary to accumulate in each light particle a great number of diverse modifications, often very difficult to reconcile with each other.

On the system of waves, the infinite variety of rays of different colours which compose white light proceed quite simply from the difference of wavelength, like the different musical tones from that of sound waves. On the Newtonian theory one cannot attribute this diversity of colours or of sensations produced on the organ of sight to difference of mass or of initial velocity of the light particles, for it would result in dispersion always being proportional to refraction, and experiment proves the contrary. Now it must necessarily be admitted that the particles of the rays of different colours are not of the same nature. There must be then as many different light particles as there are colours, of diverse shades, in the solar spectrum.

After having explained reflection and refraction by the action of repulsive and attractive forces emanating from the surface of a body, Newton, to account for the phenomena of the coloured rings, imagined, in the light particles, fits of easy reflection and of easy transmission, returning periodically at equal intervals. It was natural to suppose that these intervals, like the speed of light, were always the same in the same media, and that, in consequence, under more oblique incidence, the diameter of the rings must diminish, the distance travelled being increased. Experiment shows, on the contrary, that the diameter of the rings increases with obliquity of incidence, and Newton was obliged to conclude that the fits increased in length and to a greater degree than the distance travelled....

Thus the system of emission sufficed so little to explain the phenomena, that each new phenomenon necessitated a new hypothesis....

Not only is the hypothesis of fits improbable by its complication, and difficult to reconcile, in its consequences, with the facts, but it does not even suffice to explain the phenomenon of the coloured rings, for which it was conceived....

In the theory of waves this principle is a consequence of the fundamental hypothesis. One conceives in effect that, when two systems of light waves tend to produce opposed movements at the same point of space, they must mutually enfeeble themselves, and even destroy each other completely if the two impulsions are equal, and that the oscillations must be additive, on the other hand, when they are executed in the same phase. The intensity of the light will depend therefore on the respective positions of the two systems of waves, or, what comes to the same thing, on the difference in the distances travelled, when they emanate from a common source.

CHAPTER 13

THE CONSERVATION AND DISSIPATION
OF ENERGY

47. From JAMES JOULE's lecture, *On Matter, Living Force and Heat* (1847)

[In this popular lecture Joule states his views on the conservation of energy and on the nature of heat. His term 'living force' represents the energy of motion of a body—what is now known as kinetic energy.]

You will at once perceive that the living force of which we have been speaking is one of the most important qualities with which matter can be endowed, and, as such, that it would be absurd to suppose that it can be destroyed, or even lessened, without producing the equivalent of attraction through a given distance of which we have been speaking. You will therefore be surprised to hear that until very recently the universal opinion has been that living force could be absolutely and irrevocably destroyed at any one's option. Thus, when a weight falls to the ground, it has been generally supposed that its living force is absolutely annihilated, and that the labour which may have been expended in raising it to the elevation from which it fell has been entirely thrown away and wasted, without the production of any permanent effect whatever. We might reason, *a priori*, that such absolute destruction of living force cannot possibly take place, because it is manifestly absurd to suppose that the powers with which God has endowed matter can be destroyed any more than that they can be created by man's agency. . . .

How comes it to pass that, though in almost all natural phenomena we witness the arrest of motion and the apparent destruction of living force, we find that no waste or loss of living force has actually occurred? Experiment has enabled us to answer these questions in a satisfactory manner; for it has shown that, wherever living force is apparently destroyed, an equivalent is produced which in process of time may be reconverted into living

96

force. This equivalent is heat. Experiment has shown that wherever living force is apparently destroyed or absorbed, heat is produced. The most frequent way in which living force is thus converted into heat is by means of friction. Wood rubbed against wood or against any hard body, metal rubbed against metal or against any other body—in short, all bodies, solid or even liquid, rubbed against each other are invariably heated, sometimes even so far as to become red-hot. In all these instances the quantity of heat produced is invariably in proportion to the exertion employed in rubbing the bodies together—that is, to the living force absorbed. By fifteen or twenty smart and quick strokes of a hammer on the end of an iron rod of about a quarter of an inch in diameter placed upon an anvil an expert blacksmith will render that end of the iron visibly red-hot. Here heat is produced by the absorption of the living force of the descending hammer in the soft iron; which is proved to be the case from the fact that the iron cannot be heated if it be rendered hard and elastic, so as to transfer the living force of the hammer to the anvil.

The general rule, then, is, that wherever living force is apparently destroyed, whether by percussion, friction, or any similar means, an exact equivalent of heat is restored. The converse of this proposition is also true, namely, that heat cannot be lessened or absorbed without the production of living force, or its equivalent attraction through space. Thus, for instance, in the steam-engine it will be found that the power gained is at the expense of the heat of the fire,—that is, that the heat occasioned by the combustion of the coal would have been greater had a part of it not been absorbed in producing and maintaining the living force of the machinery. It is right, however, to observe that this has not as yet been demonstrated by experiment. But there is no room to doubt that experiment would prove the correctness of what I have said; for I have myself proved that a conversion of heat into living force takes place in the expansion of air, which is analogous to the expansion of steam in the cylinder of the steam-engine. But the most convincing proof of the conversion of heat into living force has been derived from my experiments with the electro-magnetic engine, a machine composed of magnets and bars of iron set in motion by an electrical battery. I have proved by actual experiment that, in exact proportion to the force with which this machine works, heat is abstracted from the electrical battery.

You see, therefore, that living force may be converted into heat, and that heat may be converted into living force, or its equivalent attraction through space. All three, therefore—namely, heat, living force, and attraction through space (to which I might also add light, were it consistent with the scope of the present lecture)—are mutually convertible into one another. In these conversions nothing is lost. The same quantity of heat will always be converted into the same quantity of living force. We can therefore express the equivalency in definite language applicable at all times and under all circumstances. Thus the attraction of 817 lb. through the space of one foot is equivalent to, and convertible into, the living force possessed by a body of the same weight of 817 lb. when moving with the velocity of eight feet per second, and this living force is again convertible into the quantity of heat which can increase the temperature of one pound of water by one degree Fahrenheit. The knowledge of the equivalency of heat to mechanical power is of great value in solving a great number of interesting and important questions. In the case of the steam-engine, by ascertaining the quantity of heat produced by the combustion of coal, we can find out how much of it is converted into mechanical power, and thus come to a conclusion how far the steam-engine is susceptible of further improvements. Calculations made upon this principle have shown that at least ten times as much power might be produced as is now obtained by the combustion of coal. Another interesting conclusion is, that the animal frame, though destined to fulfil so many other ends, is as a machine more perfect than the best contrived steam-engine—that is, is capable of more work with the same expenditure of fuel.

48. From J. R. MAYER's paper, *Remarks on the Forces of Inorganic Nature* (1842)

[On this paper rests Mayer's claim to priority in the enunciation of the principle of the conservation of energy and in the determination of the mechanical equivalent of heat. The following extract is the concluding part of the paper. Mayer's term 'falling force' represents the work a body is capable of performing in falling—what we should now call its potential energy.]

The natural connexion existing between falling force, motion, and heat may be conceived of as follows. We know that heat makes its appearance when the separate particles of a body approach nearer to each other: condensation produces heat. And what applies to the smallest particles of matter, and the smallest intervals between them, must also apply to large masses and to measurable distances. The falling of a weight is a real diminution of the bulk of the earth, and must therefore without doubt be related to the quantity of heat thereby developed; this quantity of heat must be proportional to the greatness of the weight and its distance from the ground. From this point of view we are very easily led to the equations between falling force, motion, and heat, that have already been discussed. . . .

If falling force and motion are equivalent to heat, heat must also naturally be equivalent to motion and falling force. Just as heat appears as an effect of the diminution of bulk and of the cessation of motion, so also does heat disappear as a cause when its effects are produced in the shape of motion, expansion, or raising of weight. *116500*

In water-mills, the continual diminution in bulk which the earth undergoes, owing to the fall of the water, gives rise to motion, which afterwards disappears again, calling forth unceasingly a great quantity of heat; and inversely, the steam-engine serves to decompose heat again into motion or the raising of weights. A locomotive engine with its train may be compared to a distilling apparatus; the heat applied under the boiler passes off as motion, and this is deposited again as heat at the axles of the wheels.

We will close our disquisition, the propositions of which have resulted as necessary consequences from the principle 'causa aequat effectum', and which are in accordance with all the phenomena of Nature, with a practical deduction. The solution of the equations subsisting between falling force and motion requires that the space fallen through in a given time, e.g. the first second, should be experimentally determined; in like manner, the solution of the equations subsisting between falling force and motion on the one hand and heat on the other, requires an answer to the question, how great is the quantity of heat which corresponds to a given quantity of motion or falling force? For instance, we must ascertain how high a given weight requires to be raised above the ground in order that its falling force may be equivalent to the raising of the temperature of an equal weight of water from 0° to 1° C. The attempt to show that such an equation is the expression of a physical truth may be regarded as the substance of the foregoing remarks.

By applying the principles that have been set forth to the relations subsisting between the temperature and the volume of gases, we find that the sinking of a mercury column by which a gas is compressed is equivalent to the quantity of heat set free by the compression; and hence it follows, the ratio between the capacity for heat of air under constant pressure and its capacity under constant volume being taken as $= 1\cdot421$, that the warming of a given weight of water from $0°$ to $1°$ C. corresponds to the fall of an equal weight from the height of about 365 metres.

If we compare with this result the working of our best steam-engines, we see how small a part only of the heat applied under the boiler is really transformed into motion or the raising of weights; and this may serve as justification for the attempts at the profitable production of motion by some other method than the expenditure of the chemical difference between carbon and oxygen—more particularly by the transformation into motion of electricity obtained by chemical means.

49. From SADI CARNOT'S *Reflections on the Motive Power of Heat* (1824)

[Carnot explains that the most convenient form of heat engine is one utilising a gas or vapour such as steam, and that its efficiency depends on the difference in temperature of the steam as it enters and leaves the engine.]

The elastic fluids, gases or vapours, are the means really adapted to the development of the motive power of heat. They combine all the conditions necessary to fulfil this office. They are easy to compress; they can be almost infinitely expanded; variations of volume occasion in them great changes of temperature; and, lastly, they are very mobile, easy to heat and to cool, easy to transport from one place to another, which enables them to produce rapidly the desired effects. We can easily conceive a multitude of machines fitted to develop the motive power of heat through the use of elastic fluids; but in whatever way we look at it, we should not lose sight of the following principles:

1. The temperature of the fluid should be made as high as

possible, in order to obtain a great fall of caloric, and consequently a large production of motive power.

2. For the same reason the cooling should be carried as far as possible.

3. It should be so arranged that the passage of the elastic fluid from the highest to the lowest temperature should be due to increase of volume; that is, it should be so arranged that the cooling of the gas should occur spontaneously as the effect of rarefaction. The limits of the temperature to which it is possible to bring the fluid primarily, are simply the limits of the temperature obtainable by combustion; they are very high.

The limits of cooling are found in the temperature of the coldest body of which we can easily and freely make use; this body is usually the water of the locality....

It is seldom that in steam-engines the elastic fluid is produced under a higher pressure than six atmospheres—a pressure corresponding to about 160° Centigrade, and it is seldom that condensation takes place at a temperature much under 40°. The fall of caloric from 160° to 40° is 120°....

Coal being capable of producing, by its combustion, a temperature higher than 1000°, and the cold water, which is generally used in our climate, being at about 10°, we can easily procure a fall of caloric of 1000°, and of this only 120° are utilized by steam-engines. Even these 120° are not wholly utilized. There is always considerable loss due to useless re-establishments of equilibrium in the caloric.

It is easy to see the advantages possessed by high-pressure machines over those of lower pressure. This superiority lies essentially in the power of utilizing a greater fall of caloric. The steam produced under a higher pressure is found also at a higher temperature, and as, further, the temperature of condensation remains always about the same, it is evident that the fall of caloric is more considerable.

CHAPTER 14

FIELD PHYSICS

50. MICHAEL FARADAY on Lines of Force and the Field

[In these passages, taken from *Experimental Researches in Electricity*, vol. III, Faraday discusses lines of magnetic force, the idea that a body extends as far as its lines of gravitational force, and that light consists of vibrations in lines of force.]

On a former occasion certain lines about a bar-magnet were described and defined (being those which are depicted to the eye by the use of iron filings sprinkled in the neighbourhood of the magnet), and were recommended as expressing accurately the nature, condition, direction, and amount of the force in any given region either within or outside of the bar. At that time the lines were considered in the abstract. Without departing from or unsettling anything then said, the inquiry is now entered upon of the possible and probable *physical existence* of such lines....

Many powers act manifestly at a distance; their physical nature is incomprehensible to us: still we may learn much that is real and positive about them, and amongst other things something of the condition of the space between the body acting and that acted upon, or between the two mutually acting bodies. Such powers are presented to us by the phaenomena of gravity, light, electricity, magnetism, &c. These when examined will be found to present remarkable differences in relation to their respective lines of forces; and at the same time that they establish the existence of real physical lines in some cases, will facilitate the consideration of the question as applied especially to magnetism....

In this view of a magnet, the medium or space around it is as essential as the magnet itself, being a part of the true and complete magnetic system. There are numerous experimental results which show us that the relation of the lines to the surrounding space can be varied by occupying it with different substances; just as the relation of a ray of light to the space through which it passes can be varied by the presence of different bodies made to occupy that space, or as the lines of electric force are affected by the media through which either induction or conduction takes place. This variation in regard to the magnetic power may be considered as depending upon the aptitude which the surrounding space has to effect the mutual relation of the two external polarities, or to carry onwards the physical line of force....

You are aware of the speculation which I some time since uttered respecting that view of the nature of matter which considers its ultimate atoms as centres of force, and not as so many little bodies surrounded by forces, the bodies being considered in the abstract as independent of the forces and capable of existing without them. In the latter view, these little particles have a definite form and a certain limited size; in the former view such is not the case, for that which represents size may be considered as extending to any distance to which the lines of force of the particle extend: the particle indeed is supposed to exist only by these forces, and where they are it is. The consideration of matter under this view gradually led me to look at the lines of force as being perhaps the seat of the vibrations of radiant phaenomena....

The view which I am so bold as to put forth considers, therefore, radiation as a high species of vibration in the lines of force which are known to connect particles and also masses of matter together. It endeavours to dismiss the æther, but not the vibrations. The kind of vibration which, I believe, can alone account for the wonderful, varied, and beautiful phaenomena of polarization, is not the same as that which occurs on the surface of disturbed water, or the waves of sound in gases or liquids, for the vibrations in these cases are direct, or to and from the centre of action, whereas the former are lateral. It seems to me, that the resultant of two or more lines of force is in an apt condition for that action which may be considered as equivalent to a *lateral* vibration; whereas a uniform medium, like the æther, does not appear apt, or more apt than air or water.

The occurrence of a change at one end of a line of force easily suggests a consequent change at the other. The propagation of light, and therefore probably of all radiant action, occupies *time*; and, that a vibration of the line of force should account for the phaenomena of radiation, it is necessary that such vibration should occupy time also. I am not aware whether there are any data by which it has been, or could be ascertained whether such a power as gravitation acts without occupying time, or whether lines of force being already in existence, such a lateral disturbance of them at one end as I have suggested above, would require time, or must of necessity be felt instantly at the other end.

51. From JAMES CLERK MAXWELL's *A Treatise on Electricity and Magnetism* (1873)

As I proceeded with the study of Faraday, I perceived that his method of conceiving the phenomena was also a mathematical one, though not exhibited in the conventional form of mathematical symbols. I found also that these methods were capable of being expressed in the ordinary mathematical forms, and thus compared with those of the professed mathematicians.

For instance, Faraday, in his mind's eye, saw lines of force traversing all space where the mathematicians saw centres of force attracting at a distance: Faraday saw a medium where they saw nothing but distance: Faraday sought the seat of the phenomena in real actions going on in the medium, they were satisfied that they had found it in a power of action at a distance impressed on the electric fluids....

Great progress has been made in electrical science, chiefly in Germany, by cultivators of the theory of action at a distance.... The great success which these eminent men have attained in the application of mathematics to electrical phenomena, gives, as is natural, additional weight to their theoretical speculations, so that those who, as students of electricity, turn to them as the greatest authorities in mathematical electricity, would probably imbibe, along with their mathematical methods, their physical hypotheses.

These physical hypotheses, however, are entirely alien from the way of looking at things which I adopt, and one object which I have in view is that some of those who wish to study electricity may, by reading this treatise, come to see that there is another way of treating the subject, which is no less fitted to explain the phenomena, and which, though it may appear less definite, corresponds, as I think, more faithfully with our actual knowledge, both in what it affirms and in what it leaves undecided.

52. From HEINRICH HERTZ'S *Electric Waves* (1892)

[Hertz was led to the discovery of electromagnetic waves by Maxwell's theory. His attitude to the theory was that of the present day, that it merely expresses relations between phenomena.]

And now, to be more precise, what is it that we call the Faraday–Maxwell theory? Maxwell has left us as the result of his mature thought a great treatise on Electricity and Magnetism; it might therefore be said that Maxwell's theory is the one which is propounded in that work. But such an answer will scarcely be regarded as satisfactory by all scientific men who have considered the question closely. Many a man has thrown himself with zeal into the study of Maxwell's work, and, even when he has not stumbled upon unwonted mathematical difficulties, has nevertheless been compelled to abandon the hope of forming for himself an altogether consistent conception of Maxwell's ideas. I have fared no better myself. Notwithstanding the greatest admiration for Maxwell's mathematical conceptions, I have not always felt quite certain of having grasped the physical significance of his statements. . . .

To the question, 'What is Maxwell's theory?' I know of no shorter or more definite answer than the following:—Maxwell's theory is Maxwell's system of equations. Every theory which leads to the same system of equations, and therefore comprises the same possible phenomena, I would consider as being a form or special case of Maxwell's theory; every theory which leads to different equations, and therefore to different possible phenomena, is a different theory. . . .

I have endeavoured to avoid from the beginning the introduction of any conceptions which are foreign to this standpoint and which might afterwards have to be removed. I have further endeavoured in the exposition to limit as far as possible the number of those conceptions which are arbitrarily introduced by us, and only to admit such elements as cannot be removed or altered without at the same time altering possible experimental results. It is true, that in consequence of these endeavours, the theory acquires a very abstract and colourless appearance. It is not particularly pleasing to hear general statements made about 'directed changes of state', where we used to have placed before our eyes pictures of electrified atoms. It is not particularly satisfactory to see equations set forth as direct results of observation and experiment, where we used to get long mathematical deductions as apparent proofs of them. Nevertheless, I believe that we cannot, without deceiving ourselves, extract much more from known facts than is asserted in the papers referred to. If we wish to lend more colour to the theory, there is nothing to prevent us from supplementing all this and aiding our powers of imagination by concrete representations of the various conceptions as to the nature of electric polarisation, the electric current, etc. But scientific accuracy requires of us that we should in no wise confuse the simple and homely figure, as it is presented to us by nature, with the gay garment which we use to clothe it. Of our own free will we can make no change whatever in the form of the one, but the cut and colour of the other we can choose as we please.

CHAPTER 15

THE RISE OF ORGANIC CHEMISTRY

53. From a paper by JEAN DUMAS and JUSTUS VON LIEBIG,
Note on the present state of Organic Chemistry (1837)

[The first glimpse of the structure of organic chemistry, contrasted with the
well-organised field of inorganic chemistry, is described in this somewhat
sanguine account of the theory of organic radicals, which was read before
the French Academy of Sciences.]

Scarcely sixty years have elapsed since the memorable epoch
when there appeared, in the midst of this assembly, the first,
fertile attempts at chemical theory which we owe to the genius
of Lavoisier. This short space of time has sufficed for the most
profound problems of inorganic chemistry to be basically considered and we are convinced that this branch of our knowledge
possesses nearly all the fundamental ideas required to deal with
the means of observation at its command.

Not only is this an incontestable fact but it is one which is
easily explained. Inorganic chemistry is concerned with the
account of elements, with their binary combinations and their
combinations into salts. Now elements can be classified in
several, very obvious groups such that if one studies carefully
the properties of one member of a group, one can almost always
predict the properties of its neighbours. The study of oxygen
informs us about sulphur; that of chlorine suffices to initiate us
into the smallest details of the properties of iodine, etc.

Thus this task, which appeared at first beyond human power,
for it was no less than the study and analysis of thousands of
substances of very different appearance and properties, has nevertheless been accomplished in less than half a century and there
remain only gaps, here and there, to be filled....

But how can we apply, with a like success, such ideas to organic
chemistry? There we meet as many kinds of compounds as in
inorganic chemistry, and they are no less diverse. Yet there, in

place of fifty-four elements, we find scarcely more than three or four in the great number of known compounds. In short, how, with the aid of the laws of inorganic chemistry, can we explain and classify the very various substances that we obtain from organic bodies, which are nearly all formed of carbon, hydrogen and oxygen only, to which is sometimes added nitrogen?

This is a great and noble problem of natural philosophy, a problem calculated to excite to the highest degree the emulation of chemists; for once resolved, the finest triumphs are promised to science. The mysteries of growth and the mysteries of animal life will unveil themselves before our eyes; we shall seize the clue to all those modifications of matter, so speedy, so sudden and so strange, which take place in animals and plants; nay more, we shall find the means of imitating them in our laboratories.

Well, we do not fear to assert, and it is not a light assertion, that this great and noble problem is today resolved; there remains only the unravelling of all the consequences which its solution entails. . . .

In fact to produce with three or four elements combinations as varied, perhaps more varied, than those which compose the whole inorganic realm, nature has taken a course as simple as unexpected; for with the elements she has made compounds which possess all the properties of the elements themselves.

There lies the whole secret of organic chemistry, we are convinced.

Thus organic chemistry possesses its own elements, which play the role of chlorine or oxygen in inorganic chemistry or, on the other hand, the role of the metals. Cyanogen, amide, benzoyl, the radicals of ammonia, of the fatty substances, of the alcohols and of analogous substances, these are the real elements with which organic chemistry operates and not the actual elements, carbon, hydrogen, oxygen and nitrogen, which appear only when all trace of their organic origin has disappeared.

For us, inorganic chemistry embraces all substances which result from the direct combination of the actual elements.

Organic chemistry, on the other hand, must deal with all substances formed from compounds functioning like the elements.

In inorganic chemistry the radicals are simple; in organic chemistry the radicals are complex. There lies the whole difference. . . .

To discover these radicals, to study them, to characterize them, such has been, for ten years, our daily endeavour. Animated by the same hope, traversing the same route, making use of the same means, it was rarely that we did not study simultaneously the same substances, or substances closely allied, and that we did not regard the facts, which presented themselves to us, from the same point of view. Sometimes, nevertheless, our opinions appeared to diverge and then, both of us carried away by the heat of the combat that we waged with nature, there sprang up between us discussions, the keenness of which we both regret. Who could deny, however, the usefulness and necessity of these discussions? Who could tell how many beautiful researches they have created, and how many they will still create?

54. From EDWARD FRANKLAND'S paper, *On a New Series of Organic Bodies containing Metals* (1852)

[Frankland's paper gave the first clear descriptionof the concept of valency, i.e. the combining-power of an element.]

When the formulae of inorganic chemical compounds are considered, even a superficial observer is impressed with the general symmetry of their construction. The compounds of nitrogen, phosphorus, antimony, and arsenic, especially, exhibit the tendency of these elements to form compounds containing 3 or 5 atoms of other elements; and it is in these proportions that their affinities are best satisfied; thus in the ternal group we have NO_3, NH_3, NI_3, NS_3, PO_3, PH_3, PCl_3, SbO_3, SbH_3, $SbCl_3$, AsO_3, AsH_3, $AsCl_3$, etc.; and in the five-atom group, NO_5, NH_4O, NH_4I, PO_5, PH_4I, etc. Without offering any hypothesis regarding the cause of this symmetrical grouping of atoms, *it is sufficiently evident, from the examples just given, that such a tendency or law prevails, and that, no matter what the character of the uniting atoms may be, the combining-power of the attracting element,* if I may be allowed the term, *is always satisfied by the same number of these atoms.*

55. From AUGUST KEKULÉ'S paper, *The Constitution and Metamorphoses of Chemical Compounds and the Chemical Nature of Carbon* (1858)

[Kekulé states two fundamental facts of organic chemistry, that the atom of carbon has a valency of 4 and that it can link itself to other carbon atoms.]

If we consider only the simplest compounds of carbon (marsh gas, methyl chloride, carbon tetrachloride, chloroform, carbonic acid, phosgene gas, carbon disulphide, prussic acid, etc.), we are struck by the fact that the amount of carbon, which the chemist has recognised as the least possible entering into the composition of a molecule, i.e. as the atom, always combines with four atoms of a monatomic, or two atoms of a diatomic, element; that in general the sum of the chemical units of the elements which combine with one atom of carbon is equal to 4. This leads to the view that carbon is tetratomic...for example,

$$CH_4 \qquad COCl_2 \qquad CO_2 \qquad CNH$$
$$CCl_4 \qquad \qquad \quad CS_2$$
$$CH_3Cl$$
$$CHCl_3$$

For substances which contain more atoms of carbon, we must assume that at least part of the atoms are held by the affinity of carbon and that the carbon atoms themselves are connected, so that naturally a part of the affinity of one for the other will bind an equal part of the affinity of the other....

When we make comparisons between compounds which have an equal number of carbon atoms in the molecule and which can be changed into each other by simple transformations (e.g. alcohol, ethyl chloride, aldehyde, acetic acid, glycolic acid, oxalic acid, etc.) we find that the carbon atoms are arranged in the same way and only the atoms held to the carbon framework are changed.

56. From AUGUST KEKULÉ's paper, *Studies on Aromatic Compounds* (1865)

[Kekulé announces his conception of the benzene ring, consisting of six carbon atoms linked by alternate single and double bonds or affinity units.]

If we wish to give an account of the atomistic constitution of aromatic compounds, we must take into consideration the following facts:

1. All aromatic compounds, even the simplest, are proportionally richer in carbon than the analogous compounds in the class of the fatty bodies.

2. Among the aromatic compounds, just as in the fatty bodies, there are numerous homologous substances, i.e., those whose differences of composition can be expressed by nCH_2.

3. The simplest aromatic compound contains at least six atoms of carbon.

4. All derivatives of aromatic substances show a certain family similarity; they belong collectively to the group of 'aromatic compounds'. Indeed in more drastic reactions, one part of carbon is often eliminated, but the chief product contains at least six atoms of carbon (benzene, quinone, chloranil, carbolic acid, hydroxyphenic acid, picric acid, etc.). The decomposition stops with the formation of these products if complete destruction of the organic group does not occur.

These facts obviously lead to the conclusion that in all aromatic substances there is contained one and the same atom group, or, if you like, a common nucleus which consists of six carbon atoms. Within this nucleus the carbon atoms are certainly in close combination or in more compact arrangement. To this nucleus, then, more carbon atoms can be added in the same way and according to the same laws as in the case of the fatty bodies.

We must next explain the atomic constitution of this nucleus. Now this can be done very easily by the following hypothesis, which, on the now generally accepted view that carbon is tetratomic, accounts for it in such a simple manner that further development is hardly necessary.

If many carbon atoms combine with one another then it can happen that one affinity unit of one atom binds one affinity unit of the neighbouring atom. As I have shown earlier, this explains homology and, on the whole, the constitution of the fatty bodies. We can further assume that many carbon atoms are linked together through two affinity units; we can also assume that the union occurs alternately through first one and then two affinity units. The first and the last of these views could be expressed as follows:

$$1/1, \quad 1/1, \quad 1/1, \quad 1/1 \text{ etc.}$$
$$1/1, \quad 2/2, \quad 1/1, \quad 2/2 \text{ etc.}$$

The first law of the symmetry of union of the carbon atoms explains the constitution of the fatty bodies, as already mentioned; the second leads to an explanation of the constitution of aromatic substances, or at least of the nucleus which is common to all these substances.

CHAPTER 16

EVOLUTION

57. From JEAN BAPTISTE LAMARCK'S *Zoological Philosophy* (1809)

[Lamarck states his view of the mechanism of evolution and the two laws which govern it; he proceeds to discuss individual species as examples.]

In the preceding chapter we saw that it is now an unquestionable fact that on passing along the animal scale in the opposite direction from that of nature, we discover the existence, in the groups composing this scale, of a continuous but irregular degradation in the organisation of animals, an increasing simplification in their organisation, and, lastly, a corresponding diminution in the number of their faculties.

This well-ascertained fact may throw the strongest light over the actual order followed by nature in the production of all the animals that she has brought into existence, but it does not show us why the increasing complexity of the organisation of animals from the most imperfect to the most perfect exhibits only an *irregular gradation*, in the course of which there occur numerous anomalies or deviations with a variety in which no order is apparent.

Now on seeking the reason of this strange irregularity in the increasing complexity of animal organisation, if we consider the influence that is exerted by the infinitely varied environments of all parts of the world on the general shape, structure and even organisation of these animals, all will then be clearly explained....

Now the true principle to be noted in all this is as follows:

1. Every fairly considerable and permanent alteration in the environment of any race of animals works a real alteration in the needs of that race.

2. Every change in the needs of animals necessitates new activities on their part for the satisfaction of those needs, and hence new habits.

3. Every new need, necessitating new activities for its satis-

faction, requires the animal, either to make more frequent use of some of its parts which it previously used less, and thus greatly to develop and enlarge them; or else to make use of entirely new parts, to which the needs have imperceptibly given birth by efforts of its inner feeling; this I shall shortly prove by means of known facts.

Thus to obtain a knowledge of the true causes of that great diversity of shapes and habits found in the various known animals, we must reflect that the infinitely diversified but slowly changing environment in which the animals of each race have successively been placed, has involved each of them in new needs and corresponding alterations in their habits. This is a truth which, once recognised, cannot be disputed. Now we shall easily discern how the new needs may have been satisfied, and the new habits acquired, if we pay attention to the two following laws of nature, which are always verified by observation.

FIRST LAW

In every animal which has not passed the limit of its development, a more frequent and continuous use of any organ gradually strengthens, develops and enlarges that organ, and gives it a power proportional to the length of time it has been so used; while the permanent disuse of any organ imperceptibly weakens and deteriorates it, and progressively diminishes its functional capacity, until it finally disappears.

SECOND LAW

All the acquisitions or losses wrought by nature on individuals, through the influence of the environment in which their race has long been placed, and hence through the influence of the predominant use or permanent disuse of any organ; all these are preserved by reproduction to the new individuals which arise, provided that the acquired modifications are common to both sexes, or at least to the individuals which produce the young....

...it was part of the plan of organisation of the reptiles, as of other vertebrates, to have four legs in dependence on their skeleton. Snakes ought consequently to have four legs, especially since they are by no means the last order of the reptiles and are farther from the fishes than are the batrachians (frogs, salamanders, etc.).

Snakes, however, have adopted the habit of crawling on the

ground and hiding in the grass; so that their body, as a result of continually repeated efforts at elongation for the purpose of passing through narrow spaces, has acquired a considerable length, quite out of proportion to its size. Now, legs would have been quite useless to these animals and consequently unused. Long legs would have interfered with their need of crawling, and very short legs would have been incapable of moving their body, since they could only have had four. The disuse of these parts thus became permanent in the various races of these animals, and resulted in the complete disappearance of these same parts, although legs really belong to the plan of organisation of the animals of this class....

The bird which is drawn to the water by its need of finding there the prey on which it lives, separates the digits of its feet in trying to strike the water and move about on the surface. The skin which unites these digits at their base acquires the habit of being stretched by these continually repeated separations of the digits; thus in course of time there are formed large webs which unite the digits of ducks, geese, etc., as we actually find them. In the same way efforts to swim, that is to push against the water so as to move about in it, have stretched the membranes between the digits of frogs, sea-tortoises, the otter, beaver, etc.

On the other hand, a bird which is accustomed to perch on trees and which springs from individuals all of whom had acquired this habit, necessarily has longer digits on its feet and differently shaped from those of the aquatic animals that I have just named. Its claws in time become lengthened, sharpened and curved into hooks, to clasp the branches on which the animal so often rests.

We find in the same way that the bird of the water-side which does not like swimming and yet is in need of going to the water's edge to secure its prey, is continually liable to sink in the mud. Now this bird tries to act in such a way that its body should not be immersed in the liquid, and hence makes its best efforts to stretch and lengthen its legs. The long-established habit acquired by this bird and all its race of continually stretching and lengthening its legs, results in the individuals of this race becoming raised as though on stilts, and gradually obtaining long, bare legs, denuded of feathers up to the thighs and often higher still....

Since ruminants can only use their feet for support, and have little strength in their jaws, which only obtain exercise by cutting and browsing on the grass, they can only fight by blows with their heads, attacking one another with their crowns.

In the frequent fits of anger to which the males especially are subject, the efforts of their inner feeling cause the fluids to flow more strongly towards that part of their head; in some there is hence deposited a secretion of horny matter, and in others of bony matter mixed with horny matter, which gives rise to solid protuberances: thus we have the origin of horns and antlers, with which the head of most of these animals is armed.

It is interesting to observe the result of habit in the peculiar shape and size of the giraffe (*Camelo-pardalis*): this animal, the largest of the mammals, is known to live in the interior of Africa in places where the soil is nearly always arid and barren, so that it is obliged to browse on the leaves of trees and to make constant efforts to reach them. From this habit long maintained in all its race, it has resulted that the animal's fore-legs have become longer than its hind-legs, and that its neck is lengthened to such a degree that the giraffe, without standing up on its hind legs, attains a height of six metres.

Among birds, ostriches, which have no power of flight and are raised on very long legs, probably owe their singular shape to analogous circumstances.

The effect of habit is quite as remarkable in the carnivorous mammals as in the herbivores; but it exhibits results of a different kind.

Those carnivores, for instance, which have become accustomed to climbing, or to scratching the ground for digging holes, or to tearing their prey, have been under the necessity of using the digits of their feet: now this habit has promoted the separation of their digits, and given rise to the formation of the claws with which they are armed.

But some of the carnivores are obliged to have recourse to pursuit in order to catch their prey: now some of these animals were compelled by their needs to contract the habit of tearing with their claws, which they are constantly burying deep in the body of another animal in order to lay hold of it, and then make efforts to tear out the part seized. These repeated efforts must have resulted in its claws reaching a size and curvature which would have greatly impeded them in walking or running on stony ground: in such cases the animal has been compelled to make further efforts to draw back its claws, which are so projecting and hooked as to get in its way. From this there has gradually resulted the formation of those peculiar sheaths, into which cats, tigers, lions, etc. withdraw their claws when they are not using them.

Hence we see that efforts in a given direction, when they are long sustained or habitually made by certain parts of a living body, for the satisfaction of needs established by nature or environment, cause an enlargement of these parts and the acquisition of a size and shape that they would never have obtained, if these efforts had not become the normal activities of the animals exerting them. Instances are everywhere furnished by observations on all known animals.

58. From CHARLES DARWIN'S *The Origin of Species* (1859)

[*The Origin of Species* is a sustained argument supported by a wealth of evidence. The following brief extracts are taken from different parts of the book.]

(1) From the *Introduction*

When on board H.M.S. 'Beagle', as naturalist, I was much struck with certain facts in the distribution of the organic beings inhabiting South America, and in the geological relations of the present to the past inhabitants of that continent. These facts, as will be seen in the latter chapters of this volume, seemed to throw some light on the origin of species—that mystery of mysteries, as it has been called by one of our greatest philosophers. On my return home, it occurred to me, in 1837, that something might perhaps be made out on this question by patiently accumulating and reflecting on all sorts of facts which could possibly have any bearing on it. After five years' work I allowed myself to speculate on the subject, and drew up some short notes; these I enlarged in 1844 into a sketch of the conclusions, which then seemed to me probable: from that period to the present day I have steadily pursued the same object. I hope that I may be excused for entering on these personal details, as I give them to show that I have not been hasty in coming to a decision.

(2) From Chapter I, *Variation under Domestication*

[In this chapter Darwin considers the great differences in the varieties of domestic animals and plants, produced by selection on the part of man.]

Believing that it is always best to study some special group, I have, after deliberation, taken up domestic pigeons....

Altogether at least a score of pigeons might be chosen, which, if shown to an ornithologist, and he were told that they were wild birds, would certainly be ranked by him as well-defined species. Moreover, I do not believe that any ornithologist would in this case place the English carrier, the short-faced tumbler, the runt, the barb, pouter, and fantail in the same genus; more especially as in each of these breeds several truly-inherited sub-breeds, or species, as he would call them, could be shown him.

Great as are the differences between the breeds of the pigeon, I am fully convinced that the common opinion of naturalists is correct, namely, that all are descended from the rock-pigeon....

I have seen it gravely remarked, that it was most fortunate that the strawberry began to vary just when gardeners began to attend to this plant. No doubt the strawberry had always varied since it was cultivated, but the slight varieties had been neglected. As soon, however, as gardeners picked out individual plants with slightly larger, earlier, or better fruit, and raised seedlings from them, and again picked out the best seedlings and bred from them, then (with some aid by crossing distinct species) those many admirable varieties of the strawberry were raised which have appeared during the last half-century.

(3) From Chapter III, *Struggle for Existence*

[Having discussed in Chapter II variations occurring in nature, Darwin now considers how these variations give rise to evolution.]

All these results, as we shall more fully see in the next chapter, follow from the struggle for life. Owing to this struggle, variations, however slight and from whatever cause proceeding, if they be in any degree profitable to the individuals of a species, in their infinitely complex relations to other organic beings and to their physical conditions of life, will tend to the preservation of such individuals, and will generally be inherited by the offspring. The offspring, also, will thus have a better chance of surviving, for, of the many individuals of any species which are periodically born, but a small number can survive. I have called this principle, by which each slight variation, if useful, is preserved, by the term Natural Selection, in order to mark its relation to man's power of selection....

A struggle for existence inevitably follows from the high rate at which all organic beings tend to increase. Every being, which during its natural lifetime produces several eggs or seeds, must suffer destruction during some period of its life, and during some season or occasional year, otherwise, on the principle of geometrical increase, its numbers would quickly become so inordinately great that no country could support the product. Hence, as more individuals are produced than can possibly survive, there must in every case be a struggle for existence, either one individual with another of the same species, or with the individuals of distinct species, or with the physical conditions of life....

In the case of every species, many different checks, acting at different periods of life, and during different seasons or years, probably come into play; some one check or some few being generally the most potent; but all will concur in determining the average number or even the existence of the species. In some cases it can be shown that widely-different checks act on the same species in different districts. When we look at the plants and bushes clothing an entangled bank, we are tempted to attribute their proportional numbers and kinds to what we call chance. But how false a view is this! Every one has heard that when an American forest is cut down, a very different vegetation springs up; but it has been observed that ancient Indian ruins in the Southern United States, which must formerly have been cleared of trees, now display the same beautiful diversity and proportion of kinds as in the surrounding virgin forest. What a struggle must have gone on during long centuries between the several kinds of trees, each annually scattering its seeds by the thousand; what war between insect and insect—between insects, snails, and other animals with birds and beasts of prey—all striving to increase, all feeding on each other, or on the trees, their seeds and seedlings, or on the other plants which first clothed the ground and thus checked the growth of the trees!

(4) From Chapter IV, *Natural Selection*

As man can produce, and certainly has produced, a great result by his methodical and unconscious means of selection, what may not natural selection effect?...

Under nature, the slightest differences of structure or constitution may well turn the nicely-balanced scale in the struggle for life, and so be preserved. How fleeting are the wishes and efforts of man! how short his time! and consequently how poor will be his results, compared with those accumulated by Nature during whole geological periods!...

This leads me to say a few words on what I have called Sexual Selection. This form of selection depends, not on a struggle for existence in relation to other organic beings, or to external conditions, but on a struggle between the individuals of one sex, generally the males, for the possession of the other sex. The result is not death to the unsuccessful competitor, but few or no offspring. Sexual selection is, therefore, less rigorous than natural selection. Generally, the most vigorous males, those which are best fitted for their places in nature, will leave most progeny. But in many cases, victory depends not so much on general vigour, as on having special weapons, confined to the male sex. A hornless stag or spurless cock would have a poor chance of leaving numerous offspring. Sexual selection, by always allowing the victor to breed, might surely give indomitable courage, length to the spur, and strength to the wing to strike in the spurred leg, in nearly the same manner as does the brutal cockfighter by the careful selection of his best cocks. How low in the scale of nature the law of battle descends, I know not; male alligators have been described as fighting, bellowing, and whirling round, like Indians in a war-dance, for the possession of the females; male salmons have been observed fighting all day long; male stag-beetles sometimes bear wounds from the huge mandibles of other males; the males of certain hymenopterous insects have been frequently seen by that inimitable observer M. Fabre, fighting for a particular female who sits by, an apparently unconcerned beholder of the struggle, and then retires with the conqueror....

The affinities of all the beings of the same class have sometimes been represented by a great tree. I believe this simile largely speaks the truth. The green and budding twigs may represent existing species; and those produced during former years may represent the long succession of extinct species. At each period of growth all the growing twigs have tried to branch out on all sides, and to overtop and kill the surrounding twigs and branches, in the same manner as species and groups of species have at all times overmastered other species in the great battle for life. The limbs divided into great branches, and these into lesser and lesser branches, were themselves once, when the tree was young, budding twigs; and this connection of the former and present buds by ramifying branches may well represent the classification of all extinct and living species in groups subordinate to groups. Of the many twigs which flourished when the tree was a mere bush, only two or three, now grown into great branches, yet survive and bear the other branches; so with the species which lived during long-past geological periods, very few have left living and modified descendants. From the first growth of the tree, many a limb and branch has decayed and dropped off; and these fallen branches of various sizes may represent those whole orders, families, and genera which have now no living representatives, and which are known to us only in a fossil state. As we here and there see a thin straggling branch springing from a fork low down in a tree, and which by some chance has been favoured and is still alive on its summit, so we occasionally see an animal like the Ornithorhynchus or Lepidosiren, which in some small degree connects by its affinities two large branches of life, and which has apparently been saved from fatal competition by having inhabited a protected station. As buds give rise by growth to fresh buds, and these, if vigorous, branch out and overtop on all sides many a feebler branch, so by generation I believe it has been with the great Tree of Life, which fills with its dead and broken branches the crust of the earth, and covers the surface with its ever-branching and beautiful ramifications.

(5) From Chapter VI, *Difficulties of the Theory*

Long before the reader has arrived at this part of my work, a crowd of difficulties will have occurred to him. Some of them are so serious that to this day I can hardly reflect on them without being in some degree staggered; but, to the best of my judgment, the greater number are only apparent, and those that are real are not, I think, fatal to the theory....

To suppose that the eye with all its inimitable contrivances for adjusting the focus to different distances, for admitting different amounts of light, and for the correction of spherical and chromatic aberration, could have been formed by natural selection, seems, I freely confess, absurd in the highest degree. When it was first said that the sun stood still and the world turned round, the common sense of mankind declared the doctrine false; but the old saying of *Vox populi, vox Dei*, as every philosopher knows, cannot be trusted in science. Reason tells me, that if numerous gradations from a simple and imperfect eye to one complex and perfect can be shown to exist, each grade being useful to its possessor, as is certainly the case; if further, the eye ever varies and the variations be inherited, as is likewise certainly the case; and if such variations should be useful to any animal under changing conditions of life, then the difficulty of believing that a perfect and complex eye could be formed by natural selection, though insuperable by our imagination, should not be considered as subversive of the theory....

In the great class of the Articulata, we may start from an optic nerve simply coated with pigment, the latter sometimes forming a sort of pupil, but destitute of a lens or other optical contrivance. With insects it is now known that the numerous facets on the cornea of their great compound eyes form true lenses, and that the cones include curiously modified nervous filaments. But these organs in the Articulata are so much diversified that Müller formerly made three main classes with seven subdivisions, besides a fourth main class of aggregated simple eyes.

When we reflect on these facts, here given much too briefly, with respect to the wide, diversified, and graduated range of structure in the eyes of the lower animals; and when we bear in mind how small the number of all living forms must be in comparison with those which have become extinct, the difficulty ceases to be very great in believing that natural selection may have converted the simple apparatus of an optic nerve, coated with pigment and invested by transparent membrane, into an optical instrument as perfect as is possessed by any member of the Articulate Class....

To arrive, however, at a just conclusion regarding the formation of the eye, with all its marvellous yet not absolutely perfect characters, it is indispensable that the reason should conquer the imagination; but I have felt the difficulty far too keenly to be surprised at others hesitating to extend the principle of natural selection to so startling a length. . . .

In living bodies, variation will cause the slight alterations, generation will multiply them almost infinitely, and natural selection will pick out with unerring skill each improvement. Let this process go on for millions of years, and during each year on millions of individuals of many kinds, and may we not believe that a living optical instrument might thus be formed as superior to one of glass, as the works of the Creator are to those of man?

(6) From Chapter XI, *On the Geological Succession of Organic Beings*

The extinction of species has been involved in the most gratuitous mystery. Some authors have even supposed that, as the individual has a definite length of life, so have species a definite duration. No one can have marvelled more than I have done at the extinction of species. When I found in La Plata the tooth of a horse embedded with the remains of Mastodon, Megatherium, Toxodon, and other extinct monsters, which all co-existed with still living shells at a very late geological period, I was filled with astonishment; for, seeing that the horse, since its introduction by the Spaniards into South America, has run wild over the whole country and has increased in numbers at an unparalleled rate, I asked myself what could so recently have exterminated the former horse under conditions of life apparently so favourable. But my astonishment was groundless. Professor Owen soon perceived that the tooth, though so like that of the existing horse, belonged to an extinct species. Had this horse been still living, but in some degree rare, no naturalist would have felt the least surprise at its rarity; for rarity is the attribute of a vast number of species of all classes, in all countries. If we ask ourselves why this or that species is rare, we answer that something is unfavourable in its conditions of life; but what that something is we can hardly ever tell. . . .

I have attempted to show that the geological record is extremely imperfect; that only a small portion of the globe has been geologically explored with care; that only certain classes of organic beings have been largely preserved in a fossil state; that the number both of specimens and of species, preserved in our museums, is absolutely as nothing compared with the number of generations which must have passed away even during a single formation....

He who rejects this view of the imperfection of the geological record, will rightly reject the whole theory. For he may ask in vain where are the numberless transitional links which must formerly have connected the closely allied or representative species, found in the successive stages of the same great formation? He may disbelieve in the immense intervals of time which must have elapsed between our consecutive formations.

(7) From Chapter xiv, *Mutual Affinities of Organic Beings*

We have seen that the members of the same class, independently of their habits of life, resemble each other in the general plan of their organisation. This resemblance is often expressed by the term 'unity of type'; or by saying that the several parts and organs in the different species of the class are homologous. The whole subject is included under the general term of Morphology. This is one of the most interesting departments of natural history, and may almost be said to be its very soul. What can be more curious than that the hand of a man, formed for grasping, that of a mole for digging, the leg of the horse, the paddle of the porpoise, and the wing of the bat, should all be constructed on the same pattern, and should include similar bones, in the same relative positions? How curious it is, to give a subordinate though striking instance, that the hind-feet of the kangaroo, which are so well fitted for bounding over the open plains,—those of the climbing, leaf-eating koala, equally well fitted for grasping the branches of trees,—those of the ground-dwelling, insect or root eating, bandicoots,—and those of some other Australian marsupials,— should all be constructed on the same extraordinary type, namely with the bones of the second and third digits extremely slender and enveloped within the same skin, so that they appear like a single toe furnished with two claws. Notwithstanding this similarity of pattern, it is obvious that the hind-feet of these several animals are used for as widely different purposes as it is possible to conceive....

Why should the brain be enclosed in a box composed of such numerous and such extraordinarily shaped pieces of bone, apparently representing vertebrae? As Owen has remarked, the benefit derived from the yielding of the separate pieces in the act of parturition by mammals, will by no means explain the same construction in the skulls of birds and reptiles. Why should similar bones have been created to form the wing and the leg of a bat, used as they are for such totally different purposes, namely flying and walking? Why should one crustacean, which has an extremely complex mouth formed of many parts, consequently always have fewer legs; or conversely, those with many legs have simpler mouths? Why should the sepals, petals, stamens, and pistils, in each flower, though fitted for such distinct purposes, be all constructed on the same pattern?

On the theory of natural selection, we can, to a certain extent, answer these questions. . . .

Organs or parts in this strange condition, bearing the plain stamp of inutility, are extremely common, or even general, throughout nature. It would be impossible to name one of the higher animals in which some part or other is not in a rudimentary condition. In the mammalia, for instance, the males possess rudimentary mammae; in snakes one lobe of the lungs is rudimentary; in birds the 'bastard-wing' may safely be considered as a rudimentary digit, and in some species the whole wing is so far rudimentary that it cannot be used for flight. What can be more curious than the presence of teeth in foetal whales, which when grown up have not a tooth in their heads; or the teeth, which never cut through the gums, in the upper jaws of unborn calves?

(8) From Chapter xv, *Recapitulation and Conclusion*

If then, animals and plants do vary, let it be ever so slightly or slowly, why should not variations or individual differences, which are in any way beneficial, be preserved and accumulated through natural selection, or the survival of the fittest? If man can by patience select variations useful to him, why, under changing and complex conditions of life, should not variations useful to nature's living products often arise, and be preserved or selected? What limit can be put to this power, acting during long ages and rigidly scrutinising the whole constitution, structure, and habits of each creature,—favouring the good and rejecting the bad? I can see no limit to this power, in slowly and beautifully adapting each form to the most complex relations of life. The theory of natural selection, even if we look no farther than this, seems to be in the highest degree probable.

CHAPTER 17

THE GERM THEORY OF DISEASE

59. From JOSEPH LISTER's article, *On a New Method of Treating Compound Fracture, Abscess etc., with Observations on the Conditions of Suppuration* (1867)

[This article was Lister's first published account of his antiseptic methods in surgery.]

The frequency of disastrous consequences in compound fracture [where the broken bone causes an open wound], contrasted with the complete immunity from danger to life or limb, in a simple fracture, is one of the most striking as well as melancholy facts in surgical practice. . . .

In the course of the year 1864 I was much struck with an account of the remarkable effects produced by carbolic acid upon the sewage of the town of Carlisle, the admixture of a very small proportion not only preventing all odour from the lands irrigated with the refuse material, but, as it was stated, destroying the entozoa which usually infest cattle fed upon such pastures.

My attention having for several years been much directed to the subject of suppuration, more especially in its relation to decomposition, I saw that such a powerful antiseptic was peculiarly adapted for experiments with a view to elucidating that subject, and while I was engaged in the investigation the applicability of carbolic acid for the treatment of compound fracture naturally occurred to me.

My first attempt of this kind was made in the Glasgow Royal Infirmary in March, 1865, in a case of compound fracture of the leg. It proved unsuccessful, in consequence, as I now believe, of improper management; but subsequent trials have more than realised my most sanguine anticipations. . . .

Case 1.—James G. . . ., aged eleven years, was admitted into the Glasgow Royal Infirmary on August 12th, 1865, with compound fracture of the left leg, caused by the wheel of an empty cart passing over the limb a little below its middle. The wound, which was about an inch and a half long, was close to, but not exactly over, the line of fracture of the tibia. A probe, however, could be passed beneath the integument over the seat of fracture and for some inches beyond it. Very little blood had been extravasated into the tissues.

My house-surgeon, Dr Macfee, acting under my instructions, laid a piece of lint dipped in liquid carbolic acid upon the wound, and applied lateral pasteboard splints padded with cotton wool, the limb resting on its outer side, with the knee bent. It was left undisturbed for four days, when, the boy complaining of some uneasiness, I removed the inner splint and examined the wound. It showed no signs of suppuration, but the skin in its immediate vicinity had a slight blush of redness. I now dressed the sore with lint soaked with water having a small proportion of carbolic acid diffused through it; and this was continued for five days, during which the uneasiness and the redness of the skin disappeared, the sore meanwhile furnishing no pus, although some superficial sloughs caused by the acid were separating. But the epidermis being excoriated by this dressing, I substituted for it a solution of one part of carbolic acid in ten to twenty parts of olive oil, which was used for four days, during which a small amount of imperfect pus was produced from the surface of the sore, but not a drop appeared from beneath the skin. It was now clear that there was no longer any danger of deep-seated suppuration, and simple water-dressing was employed. Cicatrisation proceeded just as in an ordinary granulating sore. At the expiration of six weeks I examined the condition of the bones, and, finding them firmly united, discarded the splints; and two days later the sore was entirely healed, so that the cure could not be said to have been at all retarded by the circumstance of the fracture being compound.

This, no doubt, was a favourable case, and might have done well under ordinary treatment. But the remarkable retardation of suppuration, and the immediate conversion of the compound fracture into a simple fracture with a superficial sore, were most encouraging facts.

60. From LOUIS PASTEUR'S paper, *Method for Preventing Rabies after Bites* (1885)

[Pasteur recounts his investigations with dogs, and first trials with human beings, of his method of inoculation against rabies.]

These facts established, this is the means of making a dog refractory to rabies in a relatively short time.

In a series of flasks, in which the air is maintained in a dry state by fragments of potash placed in the bottom of the vessel, is suspended each day a piece of rabid spinal cord, fresh from a rabbit which has died of rabies, the rabies having developed after seven days of incubation. Each day likewise there is inoculated under the skin of the dog a full Pravaz syringe of sterilized broth, in which has been dispersed a small fragment of one of these dried cords, commencing with a cord of an ordinal number sufficiently far from the operational day, to be quite sure that the cord is not at all virulent, as ascertained by preliminary experiments. On following days, the same operation is performed with more recent cords, separated by an interval of two days, until a very virulent cord is reached, placed only a day or two earlier in a flask.

The dog is then refractory to rabies. He can be inoculated with rabies virus under the skin, or even on the surface of the skull by trephining, without rabies showing itself.

By the application of this method I had fifty dogs, of every age and breed, refractory to rabies, without having encountered a single failure, when, unexpectedly, three people from Alsace presented themselves at my laboratory: Théodore Vone, a grocer from Meissengott, near Schlestadt, bitten in the arm, on the 4th July, by his own dog which had become rabid; Joseph Meister, aged nine years, bitten likewise on the 4th July, at 8 o'clock in the morning, by the same dog. This child, thrown to the ground by the dog, had many bites, in the hand, in the legs and in the thighs; some were deep, which made even walking difficult. The chief of these bites had been cauterized with carbolic acid only twelve hours after the accident, on the 4th July at 8 o'clock in the evening, by Dr Weber of Villé.

The third person, who had not been bitten, was the mother of little Joseph Meister.

At the autopsy on the dog, destroyed by his master, the stomach was found to be full of hay, straw and bits of wood. The dog was very rabid. Joseph Meister had been lifted from under him, covered with slaver and blood.

M. Vone had severe contusions in the arm but he assured me that his shirt had not been pierced by the fangs of the dog. As he had nothing to fear, I told him that he could go back to Alsace the same day, which he did. But I kept near at hand little Meister and his mother. The weekly meeting of the Academy of Sciences took place on that very 6th July; I there saw our colleague Dr Vulpian, to whom I recounted what had happened. Dr Vulpian, as well as Dr Grancher, professor at the Faculty of Medicine, were kind enough to come immediately to see little Joseph Meister to ascertain the state and the number of his wounds. He had no fewer than 14.

The advice of our learned colleague and of Dr Grancher was, by the intensity and number of the bites, Joseph Meister was almost certain to succumb to rabies. I communicated then to M. Vulpian and to M. Grancher, the new results which I had obtained in the study of rabies since the lecture which I had given at Copenhagen, one year earlier.

The death of this child seeming inevitable, I decided, not without lively and sore anxiety, as may well be imagined, to try on Joseph Meister the method which had constantly succeeded on dogs.

My fifty dogs, it is true, had not been bitten before I made them refractory to rabies, but I knew that this circumstance could be ignored because I had already obtained a state of immunity to rabies in a great number of dogs after they had been bitten. I had given testimony, this year, concerning this new and important progress to the members of the Commission for Rabies.

Consequently, on the 6th July, at 8 o'clock in the evening, sixty hours after the bites of the 4th July, and in the presence of Drs Vulpian and Grancher, an inoculation was made in a fold raised in the skin of the right hypochondrium[1] of little Meister, consisting of a half Pravaz syringe of a cord of a rabbit, which died of rabies on the 21st June, and which had been kept since then for fifteen days in a flask of dry air.

On the following days new inoculations were made, always in the hypochondria, under the conditions given in the table [shown on page 130].

[1] The abdomen immediately below the ribs.

I brought thus to 13 the number of inoculations and to 10 the number of days of treatment. I shall say later that a smaller number of inoculations would have been sufficient. But it will be understood that in this first attempt I had to proceed with a very special circumspection.

A half Pravaz syringe

						Cord of 14 days
7th July	9 a.m.	Cord of	23rd June			
7th ,,	6 p.m.	,,	25th ,,		,,	12 ,,
8th ,,	9 a.m.	,,	27th ,,		,,	11 ,,
8th ,,	6 p.m.	,,	29th ,,		,,	9 ,,
9th ,,	11 a.m.	,,	1st July		,,	8 ,,
10th ,,	11 a.m.	,,	3rd ,,		,,	7 ,,
11th ,,	11 a.m.	,,	5th ,,		,,	6 ,,
12th ,,	11 a.m.	,,	7th ,,		,,	5 ,,
13th ,,	11 a.m.	,,	9th ,,		,,	4 ,,
14th ,,	11 a.m.	,,	11th ,,		,,	3 ,,
15th ,,	11 a.m.	,,	13th ,,		,,	2 ,,
16th ,,	11 a.m.	,,	15th ,,		,,	1 ,,

Two fresh rabbits were inoculated by trephining with each different cord employed, so as to follow the states of virulence of the cords.

Observation of the rabbits enabled me to establish that the cords of 6th, 7th, 8th, 9th, 10th July were not virulent, for they did not make their rabbits rabid. The cords of 11th, 12th, 14th, 15th, 16th July were quite virulent, and the virulent matter found in them was stronger and stronger from one to the other. Rabies showed itself after seven days of incubation in rabbits of 15th and 16th July; after eight days in those of 12th to 14th; after fifteen days in those of 11th July.

In the last days, I had therefore inoculated Joseph Meister with the most virulent rabies virus, that from a dog strengthened by a host of transferences from rabbit to rabbit; this was a virus which gave rabies to rabbits after seven days of incubation, and after eight or ten days to dogs. I was justified in this procedure by what had happened with the fifty dogs of which I have spoken.

When the state of immunity is reached one can, without causing any inconvenience, inoculate the most virulent virus in any quantity whatsoever. This, it has always appeared to me, has no other effect than to consolidate the refractory state.

Joseph Meister escaped then, not only the rabies which his bites could have caused, but that with which I had inoculated him in the course of the treatment for immunity, more virulent than canine rabies.

The final, very virulent inoculation had also the advantage of limiting the period of apprehension of the consequences of the bites. If rabies could break out, it should show itself more quickly through a more virulent virus than through that of the bites.

From the middle of the month of August I looked forward with confidence to a healthy future for Joseph Meister. Today, even more, after the elapse of three months and three weeks since the accident, his health leaves nothing to be desired.

What interpretation is to be given to the new method which I have just made known for preventing rabies after bites? I have no intention of treating this question today in a complete way. I wish to confine myself to several preliminary details, calculated to make comprehensible the reason for the experiments which I pursued, with the aim of basing the ideas on the best of the possible interpretations.

Reviewing the methods of progressive attenuation of the deadly viruses and the prophylaxy that one can deduce from it; being given, moreover, the influence of the air on the attenuation, the first thought which offers itself to the mind to account for the effects of the method, is that the period of contact with dry air of the rabid cords diminishes progressively the intensity of the virulence of these cords until it is nullified.

From that one would be led to believe that the prophylactic method in question rests on the employment of a virus at first without appreciable activity, then a weak one and then ones more and more virulent.

I shall show ultimately that the facts are in disagreement with this way of looking at the matter. I shall show that the delays in the incubation period of rabies communicated, day by day, to the rabbits, as I have just described, to test the state of virulence of the cords dried in contact with the air, have an effect of impoverishing the quantity of the rabies viruses contained in these cords and not an effect of impoverishing their virulence.

Could one admit that the inoculation of a virus, of a constant virulence, would be capable of leading to the refractory state, if one employed it in quantities very small but daily increased? This is an interpretation of the method which I am studying from an experimental point of view.

One can give the new method yet another interpretation, an interpretation certainly very strange at first sight but which merits every consideration, because it is in harmony with certain results already known, presented by vital phenomena in some lower organisms, notably various pathogenic microbes.

Many microbes appear to give rise in their cultures to substances which have the property of hindering their proper development.

In 1880 I began researches to establish that the microbe of chicken cholera must produce a kind of self-poison. I have not succeeded in demonstrating the presence of such a substance; but I think now that this study must be taken up again—and I will not fail to do so—operating in the presence of pure carbon dioxide.

The microbe of swine fever can be cultivated in very different broths, but the results are often so feeble and so promptly arrested, relatively speaking, that the culture barely reveals feeble silky streaks in the middle of the nutritive medium. One would say, without hesitation, that a substance is produced which arrests the development of the microbe, whether one cultivates it in contact with the air or in a vacuum.

M. Ranlin, my former assistant, today professor at the Faculty of Lyon, in a very remarkable thesis which he upheld at Paris, on 22nd March 1870, established that the growth of *aspergillus niger* develops a substance which partially arrests the production of this mould when the nutritive medium does not contain iron salts.

Could it be that what constitutes the rabies virus is formed of two distinct substances and that, in addition to the one which is living and capable of multiplying in the nervous system, there is another, not living, having the property, when it is present in a suitable proportion, of arresting the development of the first? I shall examine experimentally, in my next Communication, with all the attention it deserves, this third interpretation of prophylaxy in rabies.

I have no need to remark in closing that the most important question to be resolved at the moment is the interval to be observed between the time of the bites and that of commencing treatment. This interval, in the case of Joseph Meister, was two and a half days. But we must expect cases where it is much longer.

Tuesday last, 20th October, with the obliging assistance of MM. Vulpian and Grancher, I had to begin treating a youth of fifteen, bitten very seriously on both hands, six whole days before.

I shall hasten to make known to the Academy what happens in this new trial.

The Academy will hear, not perhaps without emotion the account of the act of courage and of presence of mind of the youth whose treatment I undertook last Tuesday. He is a shepherd, 15 years old, by name Jean Baptiste Jupille, of Villers-Farlay (Jura) who on seeing a dog of suspicious gait and of large size hurl itself on a group of six of his small comrades, all younger than himself, rushed forward, armed with his whip, to confront the animal. The dog seized Jupille by his left hand. Jupille then threw the dog to the ground, keeping it under him, and opened its jaws with his right hand to release his left hand, not without receiving several new bites; then, with the thong of his whip, he bound its muzzle and, seizing one of his sabots, beat it to death.

61. From ROBERT KOCH'S lecture, *On Bacteriology and its Results* (1890)

[Koch discusses the creation of medical bacteriology and states his three rules for identifying a specific micro-organism as the cause of a specific disease.]

Bacteriology is a very young science—at least, so far as concerns us medical men. About fifteen years ago there was little more known on the subject than that, in cases of anthrax and relapsing fever, peculiar, strange objects were found in the blood, and that the so-called vibrios occur in cases of infective diseases of wounds. No proof had then been given that these objects were the cause of the respective diseases, and, with the exception of a few investigators, who were looked upon as dreamers, people regarded them rather as curiosities than as possible causes of disease.

Indeed, any other opinion was scarcely possible, because it was not established that the organisms in question were independent and specifically connected with the diseases. Bacteria had been found in putrid fluids, more particularly in the blood of strangulated animals, which could not be distinguished from the anthrax bacillus. Some investigators even thought they were not living organisms at all, but regarded them as crystalloid bodies. Bacteria, identical with the spirillum of relapsing fever, were alleged to exist in sewage and in the mouths of healthy persons; and micrococci, the same as those which are found in cases of infective diseases of wounds, were said to exist in the healthy blood and tissues.

Indeed, with the means of experimental and optical research which were then at command, it was not possible to advance beyond this point, and matters must have remained long enough in that state had not new methods of investigation been devised, which in a moment entirely altered matters, and opened up new paths into the unexplored regions.

The most minute bacteria were rendered visible by the aid of an improved system of microscopic lenses and proper methods of using them, combined with the assistance of the aniline colours as stains; and by the use of these means the special morphology of each organism could be distinguished.

At the same time, by the employment of nutritive media, liquid or solid as required, it was rendered possible to separate the various organisms from one another, and to obtain pure cultivations, by means of which the specific characteristics of each could be ascertained with certainty.

I was soon able to show what these new methods of investigation could effect. By their aid a number of new, well-characterized pathogenic organisms were discovered, and—a thing of special importance—the causal connections between them and the associated diseases were established....

The idea that micro-organisms must be the cause of infectious diseases had already been expressed long since by a few leading men, but the majority did not accept the suggestion in a very kindly way; on the contrary, the first discoveries in this direction were regarded by them with scepticism. Hence it was all the more essential to offer irrefutable evidence at the outset that the micro-organisms found in a case of a certain disease are really its cause. At that time the objection was still rightly made that it might be merely a case of the accidental coincidence of the micro-organisms and the disease, and that the former did not act the part of dangerous parasites, but only of harmless ones, which happened to find those conditions necessary for existence in the diseased organs which were not offered to them in the healthy body. Many persons admitted, indeed, the pathogenic properties of the bacteria, but regarded it as possible that they had only been transformed into pathogenic from other harmless micro-organisms, accidentally or regularly present in the body, under the influence of the morbid process.

But if it can be proved—

> Firstly, that the parasite is found in every single case of the disease in question, and under conditions corresponding to the pathological changes and the clinical course of the disease;

> Secondly, that it occurs in no other disease as an accidental and non-pathogenic parasite;

> Thirdly, that when isolated from the body and propagated through a sufficient number of pure cultivations, it can produce the disease anew;

the microbe under these circumstances cannot be an accidental accompaniment of the disease, and no other relationship between the parasite and the disease can be conceived, except that the former is the cause of the latter.

The chain of proof has been completely provided for a number of diseases, such as anthrax, tuberculosis, erysipelas, tetanus, and several diseases of animals—in general, for almost all those diseases which are communicable.

CHAPTER 18

THE NINETEENTH CENTURY

62. From AUGUSTE COMTE's *The Positive Philosophy* (1840-2)

[Comte states the essence of his philosophy, that science is the final, positive form of knowledge.]

From the study of the development of human intelligence, in all directions, and through all times, the discovery arises of a great fundamental law, to which it is necessarily subject, and which has a solid foundation of proof, both in the facts of our organization and in our historical experience. The law is this:—that each of our leading conceptions,—each branch of our knowledge,— passes successively through three different theoretical conditions: the Theological, or fictitious; the Metaphysical, or abstract; and the Scientific, or positive. In other words, the human mind, by its nature, employs in its progress three methods of philosophizing, the character of which is essentially different, and even radically opposed: viz., the theological method, the metaphysical, and the positive. Hence arise three philosophies, or general systems of conceptions on the aggregate of phenomena, each of which excludes the others. The first is the necessary point of departure of the human understanding; and the third is its fixed and definitive state. The second is merely a state of transition.

In the theological state, the human mind, seeking the essential nature of beings, the first and final causes (the origin and purpose) of all effects,—in short, Absolute knowledge,—supposes all phenomena to be produced by the immediate action of supernatural beings.

In the metaphysical state, which is only a modification of the first, the mind supposes, instead of supernatural beings, abstract forces, veritable entities (that is, personified abstractions) inherent in all beings, and capable of producing all phenomena. What is called the explanation of phenomena is, in this stage, a mere reference of each to its proper entity.

In the final, the positive state, the mind has given over the vain search after absolute notions, the origin and destination of the universe, and the causes of phenomena, and applies itself to the study of their laws,—that is, their invariable relations of succession and resemblance. Reasoning and observation, duly combined, are the means of this knowledge. What is now understood when we speak of an explanation of facts is simply the establishment of a connection between single phenomena and some general facts, the number of which continually diminishes with the progress of science.

63. From JOHN TYNDALL'S *Apology for the Belfast Address* (1874)

[Tyndall's outspoken advocacy of materialism, in his address as President of the British Association for the Advancement of Science at Belfast in August 1874, called forth a chorus of denunciation, particularly in the religious Press. The following is part of his reply.]

The expression to which the most violent exception has been taken is this: 'Abandoning all disguise, the confession I feel bound to make before you is, that I prolong the vision backward across the boundary of the experimental evidence, and discern in that Matter which we, in our ignorance, and notwithstanding our professed reverence for its Creator, have hitherto covered with opprobrium, the promise and potency of every form and quality of life.' To call it a 'chorus of dissent', as my Catholic critic does, is a mild way of describing the storm of opprobrium with which this statement has been assailed. But the first blast of passion being past, I hope I may again ask my opponents to consent to reason. First of all, I am blamed for crossing the boundary of the experimental evidence. This, I reply, is the habitual action of the scientific mind—at least of that portion of it which applies itself to physical investigation....

The course of life upon earth, as far as Science can see, has been one of amelioration—a steady advance on the whole from the lower to the higher. The continued effort of animated nature is to improve its condition and raise itself to a loftier level. In man improvement and amelioration depend largely upon the growth of conscious knowledge, by which the errors of ignorance are continually moulted, and truth is organised. It is the advance of knowledge that has given a materialistic colour to the philosophy of this age. Materialism is therefore not a thing to be mourned over, but to be honestly considered—accepted if it be wholly true, rejected if it be wholly false, wisely sifted and turned to account if it embrace a mixture of truth and error. Of late years the study of the nervous system, and its relation to thought and feeling, have profoundly occupied enquiring minds. It is our duty not to shirk—it ought rather to be our privilege to accept—the established results of such enquiries, for here assuredly our ultimate weal depends upon our loyalty to the truth. Instructed as to the control which the nervous system exercises over man's moral and intellectual nature, we shall be better prepared, not only to mend their manifold defects, but also to strenghten and purify both. Is mind degraded by this recognition of its dependence? Assuredly not. Matter, on the contrary, is raised to the level it ought to occupy, and from which timid ignorance would remove it.

64. From T. H. HUXLEY's lecture, *On the Physical Basis of Life* (1868)

[In this passage Huxley attacks materialism, which 'weighs like a nightmare' on his contemporaries.]

I have endeavoured, in the first part of this discourse, to give you a conception of the direction towards which modern physiology is tending; and I ask you, what is the difference between the conception of life as the product of a certain disposition of material molecules, and the old notion of an Archaeus governing and directing blind matter within each living body, except this— that here, as elsewhere, matter and law have devoured spirit and spontaneity? And as surely as every future grows out of past and present, so will the physiology of the future gradually extend the realm of matter and law until it is co-extensive with knowledge, with feeling, and with action.

The consciousness of this great truth weighs like a nightmare, I believe, upon many of the best minds of these days. They watch what they conceive to be the progress of materialism, in such fear and powerless anger as a savage feels, when, during an eclipse, the great shadow creeps over the face of the sun. The advancing tide of matter threatens to drown their souls; the tightening grasp of law impedes their freedom; they are alarmed lest man's moral nature be debased by the increase of his wisdom.

If the 'New Philosophy' be worthy of the reprobation with which it is visited, I confess their fears seem to me to be well founded. While, on the contrary, could David Hume be consulted, I think he would smile at their perplexities, and chide them for doing even as the heathen, and falling down in terror before the hideous idols their own hands have raised.

For, after all, what do we know of this terrible 'matter', except as a name for the unknown and hypothetical cause of states of our own consciousness? And what do we know of that 'spirit' over whose threatened extinction by matter a great lamentation is arising, like that which was heard at the death of Pan, except that it is also a name for an unknown and hypothetical cause, or condition, of states of consciousness? In other words, matter and spirit are but names for the imaginary substrata of groups of natural phaenomena.

And what is the dire necessity and 'iron' law under which men groan? Truly, most gratuitously invented bugbears. I suppose if there be an 'iron' law, it is that of gravitation; and if there be a physical necessity, it is that a stone, unsupported, must fall to the ground. But what is all we really know, and can know, about the latter phaenomenon? Simply, that, in all human experience, stones have fallen to the ground under these conditions; that we have not the smallest reason for believing that any stone so circumstanced will not fall to the ground; and that we have, on the contrary, every reason to believe that it will so fall. It is very convenient to indicate that all the conditions of belief have been fulfilled in this case, by calling the statement that unsupported stones will fall to the ground, 'a law of nature'. But when, as commonly happens, we change *will* into *must*, we introduce an idea of necessity which most assuredly does not lie in the observed facts, and has no warranty that I can discover elsewhere. For my part, I utterly repudiate and anathematize the intruder. Fact I know; and Law I know; but what is this Necessity, save an empty shadow of my own mind's throwing?

But, if it is certain that we can have no knowledge of the nature of either matter or spirit, and that the notion of necessity is something illegitimately thrust into the perfectly legitimate conception of law, the materialistic position that there is nothing in the world but matter, force, and necessity, is as utterly devoid of justification as the most baseless of theological dogmas.

65. From ERNST HAECKEL'S *The Riddle of the Universe* (1899)

[Haeckel's book is not a great work but it typifies a school of thought at the end of the nineteenth century.]

The supreme and all-pervading law of nature, the true and only cosmological law, is, in my opinion, *the law of substance*; its discovery and establishment is the greatest intellectual triumph of the nineteenth century, in the sense that all other known laws of nature are subordinate to it. Under the name of 'law of substance' we embrace two supreme laws of different origin and age—the older is the chemical law of the 'conservation of matter', and the younger is the physical law of the 'conservation of energy'. It will be self-evident to many readers, and it is acknowledged by most of the scientific men of the day, that these two great laws are essentially inseparable....

Once modern physics had established the law of substance as far as the simpler relations of inorganic bodies are concerned, physiology took up the story, and proved its application to the entire province of the organic world. It showed that all the vital activities of the organism—without exception—are based on a constant 'reciprocity of force' and a correlative change of material, or metabolism, just as much as the simplest processes in 'lifeless' bodies. Not only the growth and the nutrition of plants and animals, but even their functions of sensation and movement, their sense-action and psychic life, depend on the conversion of potential into kinetic energy, and *vice versa*. This supreme law dominates also those elaborate performances of the nervous system which we call, in the higher animals and man, 'the action of the mind'.

Our monistic view, that the great cosmic law applies throughout the whole of nature, is of the highest moment. For it not only involves, on its positive side, the essential unity of the cosmos and the causal connection of all phenomena that come within our cognizance, but it also, in a negative way, marks the highest intellectual progress, in that it definitely rules out the three central dogmas of metaphysics—God, freedom, and immortality. In assigning mechanical causes to phenomena everywhere, the law of substance comes into line with the universal law of causality....

One of the most distinctive features of the expiring century is the increasing vehemence of the opposition between science and Christianity. That is both natural and inevitable. In the same proportion in which the victorious progress of modern science has surpassed all the scientific achievements of earlier ages has the untenability been proved of those mystic views which would subdue reason under the yoke of an alleged revelation; and the Christian religion belongs to that group. The more solidly modern astronomy, physics, and chemistry have established the sole dominion of inflexible natural laws in the universe at large, and modern botany, zoology, and anthropology have proved the validity of those laws in the entire kingdom of organic nature, so much the more strenuously has the Christian religion, in association with dualistic metaphysics, striven to deny the application of these natural laws in the province of the so-called 'spiritual life'—that is, in one section of the physiology of the brain.

66. From HERBERT SPENCER'S *First Principles* (1862)

[In this passage Spencer argues that Evolution is a single process of integrated differentiation, which is proceeding in all parts of the universe.]

The law of Evolution has been thus far contemplated as holding true of each order of existences, considered as a separate order. But the induction as so presented falls short of that completeness which it gains when we contemplate these several orders of existences as forming together one natural whole. While we think of Evolution as divided into astronomic, geologic, biologic, psychologic, sociologic, &c., it may seem to some extent a coincidence that the same law of metamorphosis holds throughout all its divisions. But when we recognize these divisions as mere conventional groupings, made to facilitate the arrangement and acquisition of knowledge—when we remember that the different existences with which they severally deal are component parts of one Cosmos; we see at once that there are not several kinds of Evolution having certain traits in common, but one Evolution going on everywhere after the same manner. We have repeatedly observed that while any whole is evolving, there is always going on an evolution of the parts into which it divides itself; but we have not observed that this equally holds of the totality of things, which is made up of parts within parts from the greatest down to the smallest. We know that while a physically-cohering aggregate like the human body is getting larger and taking on its general shape, each of its organs is doing the same; that while each organ is growing and becoming unlike others, there is going on a differentiation and integration of its component tissues and vessels; and that even the components of these components are severally increasing and passing into more definitely heterogeneous structures. But we have not duly remarked that while each individual is developing, the society of which he is an insignificant unit is developing too; that while the aggregate mass forming a society is integrating and becoming more definitely heterogeneous, so, too, that total aggregate, the Earth, is continuing to integrate and differentiate; that while the Earth, which in bulk is not a millionth of the Solar System, progresses towards its more concentrated structure, the Solar System similarly progresses.

So understood, Evolution becomes not one in principle only, but one in fact. There are not many metamorphoses similarly carried on, but there is a single metamorphosis universally progressing, wherever the reverse metamorphosis has not set in. In any locality, great or small, where the occupying matter acquires an appreciable individuality, or distinguishableness from other matter, there Evolution goes on; or rather, the acquirement of this appreciable individuality is the commencement of Evolution. And this holds regardless of the size of the aggregate, and regardless of its inclusion in other aggregates.

CHAPTER 19

GENETICS AND NEO-DARWINISM

67. From AUGUST WEISMANN'S *The Germ-Plasm, A Theory of Heredity* (1892)

[This extract is taken from the Preface. Weismann refers to Darwin's theory of pangenesis, that minute particles, called gemmules, are released by all parts of the body and are absorbed by the germ cells, i.e. the male sperms or female ova, thereby explaining the effect of environment on heredity. Weismann denies that such an effect can occur because the germ cells reproduce themselves from generation to generation independently of the rest of the body.]

What first struck me when I began seriously to consider the problem of heredity, some ten years ago, was the necessity for assuming the existence of a special organised and living *hereditary substance*, which in all multicellular organisms, unlike the substance composing the perishable body of the individual, is transmitted from generation to generation. This is the theory of *the continuity of the germ-plasm.* My conclusions led me to doubt the usually accepted view of the *transmission of variations acquired* by the body (soma); and further research, combined with experiments, tended more and more to strengthen my conviction that in point of fact no such transmission occurs. Meanwhile, the investigations of several distinguished biologists—in which I myself have had some share—on the process of fertilisation and conjugation, brought about a complete revolution in our previous ideas as to the meaning of this process, and further led me to see that the germ-plasm is composed of vital units, each of equal value, but differing in character, containing all the primary constituents of an individual....

All my investigations on the problem of heredity were so far only links, to be some day united into a chain which had as yet no existence. The question of the ultimate elements on which to base the theory was the very point on which I remained longest in doubt. The 'pangenesis' of Darwin, as already mentioned, seemed to me to be far too independent of facts, and even now I am of the opinion that the very hypothesis from which it derives its name is untenable. There is now scarcely any doubt that the entire conception of the production of the 'gemmules' by the body-cells, their separation from the latter, and their 'circulation', is in reality wholly imaginary. In this regard I am still quite as much opposed to Darwin's views as formerly, for I believe that all parts of the body do not contribute to produce a germ from which the new individual arises, but that on the contrary, the offspring owes its origin to a peculiar substance of extremely complicated structure, viz., the 'germ-plasm'. This substance can never be formed anew; it can only grow, multiply, and be transmitted from one generation to another.

68. From GREGOR MENDEL's paper, *Plant-Hybridisation* (1865) [The essential findings of Mendel's experiments are given in this extract.]

...The various forms of Peas selected for crossing showed differences in the length and colour of the stem; in the size and form of the leaves; in the position, colour, and size of the flowers; in the length of the flower stalk; in the colour, form, and size of the pods; in the form and size of the seeds; and in the colour of the seed-coats and of the albumen (cotyledons)....

The Forms of the Hybrids [F_1]

...In the case of each of the seven crosses the hybrid-character resembles that of one of the parental forms so closely that the other either escapes observation completely or cannot be detected with certainty. This circumstance is of great importance in the determination and classification of the forms under which the offspring of the hybrids appear. Henceforth in this paper those characters which are transmitted quite, or almost unchanged in the hybridisation, and therefore in themselves constitute the characters of the hybrid, are termed the *dominant*, and those which become latent in the process *recessive*. The expression 'recessive' has been chosen because the characters thereby designated withdraw or entirely disappear in the hybrids, but nevertheless reappear unchanged in their progeny, as will be demonstrated later on.

It was furthermore shown by the whole of the experiments that it is perfectly immaterial whether the dominant character belong to the seed-bearer or to the pollen-parent; the form of the hybrid remains identical in both cases. This interesting fact was also emphasised by Gärtner, with the remark that even the most practised expert is not in a position to determine in a hybrid which of the two parental species was the seed or the pollen plant. . . .

The First Generation (bred) from the Hybrids [F₂]

In this generation there reappear, together with the dominant characters, also the recessive ones with their peculiarities fully developed, and this occurs in the definitely expressed average proportion of three to one, so that among each four plants of this generation three display the dominant character and one the recessive. This relates without exception to all the characters which were investigated in the experiments. . . .

Expt. 1 Form of seed—From 253 hybrids 7,324 seeds were obtained in the second trial year. Among them were 5,474 round or roundish ones and 1,850 angular wrinkled ones. Therefrom the ratio 2·96 to 1 is deduced.

Expt. 2 Colour of albumen—258 plants yielded 8,023 seeds, 6,022 yellow, and 2,001 green; their ratio, therefore, is as 3·01 to 1. . . .

The Second Generation (bred) from the Hybrids [F₃]

Those forms which in the first generation (F_2) exhibit the recessive character do not further vary in the second generation (F_3) as regards this character; they remain constant in their offspring.

It is otherwise with those which possess the dominant character in the first generation (bred from the hybrids). Of these two-thirds yield offspring which display the dominant and recessive characters in the proportion of 3 to 1, and thereby show exactly the same ratio as the hybrid forms, while only one-third remains with the dominant character constant.

69. From T. H. MORGAN'S *The Theory of the Gene* (1926)

[Morgan was one of the founders of modern genetics, and in this extract he gives a concise statement of his theory of the gene.]

We are now in a position to formulate the theory of the gene. The theory states that the characters of the individual are referable to paired elements (genes) in the germinal material that are held together in a definite number of linkage groups; it states that the members of each pair of genes separate when the germ-cells mature in accordance with Mendel's first law, and in consequence each germ-cell comes to contain one set only; it states that the members belonging to different linkage groups assort independently in accordance with Mendel's second law; it states that an orderly interchange—crossing-over—also takes place, at times, between the elements in corresponding linkage groups; and it states that the frequency of crossing-over furnishes evidence of the linear order of the elements in each linkage group and of the relative position of the elements with respect to each other.

These principles, which, taken together, I have ventured to call the theory of the gene, enable us to handle problems of genetics on a strictly numerical basis, and allow us to predict, with a great deal of precision, what will occur in any given situation. In these respects the theory fulfils the requirements of a scientific theory in the fullest sense.

70. From a statement by the Praesidium of the U.S.S.R. Academy of Sciences (1948)

[In Russia the teaching of western genetics, called 'Weismannite-Morganist idealist teaching' in the following extract, is prohibited and it has been replaced by a form of neo-Lamarckism, first put forward by Michurin.]

Michurin's materialist direction in biology is the only acceptable form of science, because it is based on dialectical materialism, and on the revolutionary principle of changing Nature for the benefit of the people. Weismannite-Morganist idealist teaching is pseudo-scientific, because it is founded on the notion of the divine origin of the world and assumes eternal and unalterable scientific laws. The struggle between the two ideas has taken the form of the ideological class-struggle between socialism and capitalism on the international scale, and between the majority of Soviet scientists and a few remaining Russian scientists who have retained traces of bourgeois ideology, on a smaller scale. There is no place for compromise. Michurinism and Morgano-Weismannism cannot be reconciled.

71. From T. D. LYSENKO's address, *The Situation in Biological Science* (1948)

[Lysenko explains the Soviet view of heredity and attacks neo-Mendelism.]

Contrary to Mendelism-Morganism, with its assertion that the causes of variation in the nature of organisms are unknowable and its denial that directed changes in the nature of plants and animals are possible, I. V. Michurin's motto was: 'We cannot wait for favours from Nature; we must wrest them from her.'

His studies and investigations led I. V. Michurin to the following important conclusion: 'It is possible, with man's intervention, to *force* any form of animal or plant *to change more quickly and in a direction desirable to man.* There opens before man a broad field of activity of the greatest value to him.'

The Michurin teaching flatly rejects the fundamental principle of Mendelism-Morganism that heredity is completely independent of the plants' or animals' conditions of life. The Michurin teaching does not recognize the existence in the organism of a separate hereditary substance which is independent of the body. Changes in the heredity of an organism or in the heredity of any part of its body are the result of changes in the living body itself. And changes of the living body occur as the result of departure from the normal in the type of assimilation and dissimilation, of departure from the normal in the type of metabolism. Changes in organisms or in their separate organs or characters may not always, or not fully, be transmitted to the offspring, but changed germs of newly generated organisms always occur only as the result of changes in the body of the parent organism, as the result of direct or indirect action of the conditions of life upon the development of the organism or its separate parts, among them the sexual or vegetative germs. Changes in heredity, acquisition of new characters and their augmentation and accumulation in successive generations are always determined by the organism's conditions of life. Heredity changes and its complexity increases as the result of the accumulation of new characters and properties acquired by organisms in successive generations.

The organism and the conditions required for its life constitute a unity. Different living bodies require different environmental conditions for their development. By studying the character of these requirements we come to know the qualitative features of the nature of organisms, the qualitative features of heredity. Heredity is *the property of a living body to require definite conditions for its life and development and to respond in a definite way to various conditions....*

The representatives of Mendel-Morgan genetics are not only unable to obtain alterations of heredity in a definite direction, but categorically deny that it is possible to change heredity so that it will adequately correspond to the action of the environmental conditions. The principles of Michurin's teaching, on the other hand, tell us that it is possible to obtain changes in heredity *fully corresponding to the effect of the action of conditions of life.*

A case in point is the experiments to convert spring forms of cereal grains into winter forms, and winter forms into forms of still greater winter habit in regions of Siberia, for example, where the winters are severe. These experiments are not only of theoretical interest. They are of considerable practical value for the production of winter-hardy varieties. We already have winter forms of wheat obtained from spring forms, which are not inferior, as regards frost resistance, to the most frost-resistant varieties known in practical farming. Some are even superior.

Many experiments show that when an old established property of heredity is being eliminated, we do not at once get a fully established, solidified new heredity. In the vast majority of cases, what we get is an organism with a plastic nature, which I. V. Michurin called 'destabilized'.

Plant organisms with a 'destabilized' nature are those in which their conservatism has been eliminated, and their electivity with regard to external conditions is weakened. Instead of conservative heredity, such plants preserve, or there appears in them only a *tendency* to show some preference for certain conditions.

The nature of a plant organism may be destabilized:

1. By grafting, i.e., by uniting the tissues of plants of different breeds;

2. By bringing external conditions to bear upon it at definite moments when the organism undergoes this or that process of its development;

3. By crossbreeding, particularly of forms sharply differing in habitat or origin.

The best biologists, first and foremost I. V. Michurin, have devoted a great deal of attention to the practical value of plant organisms with destabilized heredity. Plastic plant forms with unestablished heredity, obtained by any of the enumerated methods, should be further bred from generation to generation in those conditions, the requirement of which, or adaptability to which, we want to induce and perpetuate in the given organisms.

CHAPTER 20

THE STRUCTURE OF THE ATOM

72. From J. J. THOMSON's paper, *Cathode Rays* (1897)

[In this paper J. J. Thomson describes his quantitative experiments on cathode rays and puts forward the hypothesis that the atoms of the elements are divisible and contain much smaller particles, which he terms corpuscles and which we now know as electrons.]

The experiments discussed in this paper were undertaken in the hope of gaining some information as to the nature of the Cathode Rays. The most diverse opinions are held as to these rays; according to the almost unanimous opinion of German physicists they are due to some process in the æther to which—inasmuch as in a uniform magnetic field their course is circular and not rectilinear—no phenomenon hitherto observed is analogous; another view of these rays is that, so far from being wholly ætherial, they are in fact wholly material, and that they mark the paths of particles of matter charged with negative electricity. It would seem at first sight that it ought not to be difficult to discriminate between views so different, yet experience shows that this is not the case, as amongst the physicists who have most deeply studied the subject can be found supporters of either theory.

The electrified-particle theory has for purposes of research a great advantage over the ætherial theory, since it is definite and its consequences can be predicted; with the ætherial theory it is impossible to predict what will happen under any given circumstances, as on this theory we are dealing with hitherto unobserved phenomena in the æther, of whose laws we are ignorant....

An objection very generally urged against the view that the cathode rays are negatively electrified particles, is that hitherto no deflexion of the rays has been observed under a small electrostatic force, and though the rays are deflected when they pass near electrodes connected with sources of large differences of potential, such as induction-coils or electrical machines, the deflexion in this case is regarded by the supporters of the ætherial theory as due to the discharge passing between the electrodes, and not primarily to the electrostatic field, Hertz made the rays travel between two parallel plates of metal placed inside the discharge-tube, but found that they were not deflected when the plates were connected with a battery of storage cells; on repeating this experiment I at first got the same result, but subsequent experiments showed that the absence of deflexion is due to the conductivity conferred on the rarefied gas by the cathode rays. On measuring this conductivity it was found that it diminished very rapidly as the exhaustion increased; it seemed then that on trying Hertz's experiment at very high exhaustions there might be a chance of detecting the deflexion of the cathode rays by an electrostatic force.

The apparatus used is represented in the Figure.

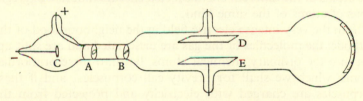

The rays from the cathode C pass through a slit in the anode A, which is a metal plug fitting tightly into the tube and connected with the earth; after passing through a second slit in another earth-connected metal plug B, they travel between two parallel aluminium plates about 5 cm. long by 2 broad and at a distance of 1·5 cm. apart; they then fall on the end of the tube and produce a narrow well-defined phosphorescent patch. A scale pasted on the outside of the tube serves to measure the deflexion of this patch. At high exhaustions the rays were deflected when the two aluminium plates were connected with the terminals of a battery of small storage-cells; the rays were depressed when the upper plate was connected with the positive, the lower with the negative pole. The deflexion was proportional to the difference of potential between the plates, and I could detect the deflexion when the potential-difference was as small as two volts. . . .

[Thomson then goes on to describe how, by measuring also the deflection of the cathode rays in the magnetic field, and by other experiments, he obtained a value of m/e, where m and e are the mass and charge respectively of each cathode ray particle.]

From these determinations we see that the value of m/e is independent of the nature of the gas, and that its value 10^{-7} is very small compared with the value 10^{-4}, which is the smallest value of this quantity previously known, and which is the value for the hydrogen ion in electrolyses.

Thus for the carriers of electricity in the cathode rays m/e is very small compared with its value in electrolyses. The smallness of m/e may be due to the smallness of m or the largeness of e, or to a combination of these two. . . .

The explanation which seems to me to account in the most simple and straightforward manner for the facts is founded on a view of the constitution of the chemical elements which has been favourably entertained by many chemists; this view is that the atoms of the different chemical elements are different aggregations of atoms of the same kind. . . .

If, in the very intense electric field in the neighbourhood of the cathode, the molecules of the gas are dissociated and are split up, not into the ordinary chemical atoms, but into these primordial atoms, which we shall for brevity call corpuscles; and if these corpuscles are charged with electricity and projected from the cathode by the electric field, they would behave exactly like the cathode rays. . . .

Thus on this view we have in the cathode rays matter in a new state, a state in which the subdivision of matter is carried very much further than in the ordinary gaseous state; a state in which all matter—that is, matter derived from different sources such as hydrogen, oxygen, etc.—is of one and the same kind; this matter being the substance from which all the chemical elements are built up.

73. From the paper by E. RUTHERFORD and F. SODDY, *The Cause and Nature of Radioactivity* (1902)

[This paper announced the startling theory that radioactivity consists of the spontaneous breaking up of atoms.]

Turning from the experimental results to their theoretical interpretation, it is necessary to first consider the generally accepted view of the nature of radioactivity. It is well established that this property is the function of the atom and not of the molecule. Uranium and thorium, to take the most definite cases, possess the property in whatever molecular condition they occur, and the former also in the elementary state. So far as the radioactivity of different compounds of different density and states of division can be compared together, the intensity of the radiation appears to depend only on the quantity of active element present. It is not at all dependent on the source from which the element is derived, or the process of purification to which it has been subjected, provided sufficient time is allowed for the equilibrium point to be reached. It is not possible to explain the phenomena by the existence of impurities associated with the radioactive elements, even if any advantage could be derived from the assumption. For these impurities must necessarily be present always to the same extent in different specimens derived from the most widely different sources, and, moreover, they must persist in unaltered amount after the most refined processes of purification. This is contrary to the accepted meaning of the term impurity.

All the most prominent workers in this subject are agreed in considering radioactivity an atomic phenomenon. M. and Mme Curie, the pioneers in the chemistry of the subject, have recently put forward their views (*Comptes Rendus*, CXXXIV, 1902, p. 85). They state that this idea underlies their whole work from the beginning and created their methods of research. M. Becquerel, the original discoverer of the property for uranium, in his announcement of the recovery of the activity of the same element after the active constituent had been removed by chemical treatment, points out the significance of the fact that uranium is giving out cathode-rays. These, according to the hypothesis of Sir William Crookes and Prof. J. J. Thomson, are *material* particles of mass one thousandth of the hydrogen atom.

Since, therefore, radioactivity is at once an atomic phenomenon and accompanied by chemical changes in which new types of matter are produced, these changes must be occurring within the atom, and the radioactive elements must be undergoing spontaneous transformation. The results that have so far been obtained, which indicate that the velocity of this reaction is unaffected by the conditions, makes it clear that the changes in question are different in character from any that have been before dealt with in chemistry. It is apparent that we are dealing with phenomena outside the sphere of known atomic forces. Radio-activity may therefore be considered as a manifestation of subatomic chemical change.

74. From ERNEST RUTHERFORD'S Bakerian lecture, *Nuclear Constitution of Atoms* (1920)

[In the first part of this extract Rutherford discusses the establishment of the nuclear theory of the atom; in the second part he describes the artificial disintegration of the nucleus of the nitrogen atom.]

Introduction.—The conception of the nuclear constitution of atoms arose initially from attempts to account for the scattering of α-particles through large angles in traversing thin sheets of matter. Taking into account the large mass and velocity of the α-particles, these large deflexions were very remarkable, and indicated that very intense electric or magnetic fields exist within the atom. To account for these results, it was found necessary to assume that the atom consists of a charged massive nucleus of dimensions very small compared with the ordinarily accepted magnitude of the diameter of the atom. This positively charged nucleus contains most of the mass of the atom, and is surrounded at a distance by a distribution of negative electrons equal in number to the resultant positive charge on the nucleus. Under these conditions, a very intense electric field exists close to the nucleus, and the large deflexion of the α-particle in an encounter with a single atom happens when the particle passes close to the nucleus. Assuming that the electric forces between the α-particle and the nucleus varied according to an inverse square law in the region close to the nucleus, the writer worked out the relations connecting the number of α-particles scattered through any angle with the charge on the nucleus and the energy of the α-particle. Under the central field of force, the α-particle describes a hyperbolic orbit round the nucleus, and the magnitude of the deflexion depends on the closeness of approach to the nucleus. From the data of scattering of α-particles then available, it was deduced that the resultant charge on the nucleus was about $\frac{1}{2}Ae$, where A is the atomic weight and e the fundamental unit of charge. Geiger and Marsden made an elaborate series of experiments to test the correctness of the theory, and confirmed the main conclusions. They found the nucleus charge was about $\frac{1}{2}Ae$, but, from the nature of the experiments, it was difficult to fix the actual value within about 20 per cent. C. G. Darwin worked out completely the deflexion of the α-particle and of the nucleus, taking into account the mass of the latter, and showed that the scattering experiments of Geiger and Marsden could be reconciled with any law of central force, except the inverse square. The nuclear constitution of the atom was thus very strongly supported by the experiments on scattering of α-rays.

Since the atom is electrically neutral, the number of external electrons surrounding the nucleus must be equal to the number of units of resultant charge on the nucleus. It should be noted that, from the consideration of the scattering of X-rays by light elements, Barkla had shown, in 1911, that the number of electrons was equal to about half the atomic weight. This was deduced from the theory of scattering of Sir J. J. Thomson, in which it was assumed that each of the external electrons in an atom acted as an independent scattering unit.

Two entirely different methods had thus given similar results with regard to the number of external electrons in the atom, but the scattering of α-rays had shown in addition that the positive charge must be concentrated on a massive nucleus of small dimensions. It was suggested by Van den Broek that the scattering of α-particles by the atoms was not inconsistent with the possibility that the charge on the nucleus was equal to the atomic number of the atom, i.e., to the number of the atom when arranged in order of increasing atomic weight. The importance of the atomic number in fixing the properties of an atom was shown by the remarkable work of Moseley on the X-ray spectra of the elements. He showed that the frequency of vibration of corresponding lines in the X-ray spectra of the elements depended on the square of a number which varies by unity in successive elements. This relation received an interpretation by supposing that the nuclear charge varied by unity in passing from atom to atom, and was given numerically by the atomic number. I can only emphasise in passing the great importance of Moseley's work, not only in fixing the number of possible elements, and the position of undetermined elements, but in showing that the properties of an atom were defined by a number which varied by unity in successive atoms. This gives a new method of regarding the periodic classification of the elements, for the atomic number, or its equivalent the nuclear charge, is of more fundamental importance than its atomic weight. ...

Long Range Particles from Nitrogen

In previous papers, I have given an account of the effects produced by close collisions of swift α-particles with light atoms of matter with the view of determining whether the nuclear structure of some of the lighter atoms could be disintegrated by the intense forces brought into play in such close collisions. Evidence was given that the passage of α-particles through dry nitrogen gives rise to swift particles which closely resembled in brilliancy of the scintillations and distance of penetration hydrogen atoms set in motion by close collision with α-particles. It was shown that these swift atoms which appeared only in dry nitrogen and not in oxygen or carbon dioxide could not be ascribed to the presence of water vapour or other hydrogen material but must arise from the collision of α-particles with nitrogen atoms. The number of such scintillations due to nitrogen was small, viz., about 1 in 12 of the corresponding number in hydrogen, but was two to three times the number of natural scintillations from the source. ...

In a previous paper I have given evidence that the long range particles observed in dry air and pure nitrogen must arise from the nitrogen atoms themselves. It is thus clear that some of the nitrogen atoms are disintegrated by their collision with swift α-particles and that swift atoms of positively charged hydrogen are expelled. It is to be inferred that the charged atom of hydrogen is one of the components of which the nucleus of nitrogen is built up.

CHAPTER 21

THE THEORY OF RELATIVITY

75. From ALBERT EINSTEIN's lecture, *The Theory of Relativity* (1921)

[The following is a brief summary of the theory of relativity in its creator's own words.]

It is a particular pleasure to me to have the privilege of speaking in the capital of the country from which the most important fundamental notions of theoretical physics have issued. I am thinking of the theory of mass motion and gravitation which Newton gave us and the concept of the electromagnetic field, by means of which Faraday and Maxwell put physics on a new basis. The theory of relativity may indeed be said to have put a sort of finishing touch to the mighty intellectual edifice of Maxwell and Lorentz, inasmuch as it seeks to extend field physics to all phenomena, gravitation included.

Turning to the theory of relativity itself, I am anxious to draw attention to the fact that this theory is not speculative in origin; it owes its invention entirely to the desire to make physical theory fit observed fact as well as possible. We have here no revolutionary act but the natural continuation of a line that can be traced through centuries. The abandonment of certain notions connected with space, time, and motion hitherto treated as fundamentals must not be regarded as arbitrary, but only as conditioned by observed facts.

The law of the constant velocity of light in empty space, which has been confirmed by the development of electrodynamics and optics, and the equal legitimacy of all inertial systems[1] (special principle of relativity), which was proved in a particularly incisive manner by Michelson's famous experiment, between them made it necessary, to begin with, that the concept of time should be

---

[1] An inertial system is one in which the laws of mechanics apply, i.e. all systems in uniform relative motion.

made relative, each inertial system being given its own special time. As this notion was developed, it became clear that the connection between immediate experience on one side and co-ordinates and time on the other had hitherto not been thought out with sufficient precision. It is in general one of the essential features of the theory of relativity that it is at pains to work out the relations between general concepts and empirical facts more precisely. The fundamental principle here is that the justification for a physical concept lies exclusively in its clear and unambiguous relation to facts that can be experienced. According to the special theory of relativity, spatial coordinates and time still have an absolute character in so far as they are directly measurable by stationary clocks and bodies. But they are relative in so far as they depend on the state of motion of the selected inertial system. According to the special theory of relativity the four-dimensional continuum formed by the union of space and time (Minkowski) retains the absolute character which, according to the earlier theory, belonged to both space and time separately. The influence of motion (relative to the coordinate system) on the form of bodies and on the motion of clocks, also the equivalence of energy and inert mass, follow from the interpretation of co-ordinates and time as products of measurement.

The general theory of relativity owes its existence in the first place to the empirical fact of the numerical equality of the inertial and gravitational mass of bodies,[1] for which fundamental fact classical mechanics provided no interpretation. Such an interpretation is arrived at by an extension of the principle of relativity to coordinate systems accelerated relatively to one another. The introduction of coordinate systems accelerated relatively to inertial systems involves the appearance of gravitational fields relative to the latter. As a result of this, the general theory of relativity, which is based on the equality of inertia and weight, provides a theory of the gravitational field.

The introduction of coordinate systems accelerated relatively to each other as equally legitimate systems, such as they appear

---

[1] The inertial mass of a body is defined in terms of the acceleration it acquires under a given force, and the gravitational mass by the attraction of other bodies, such as the earth. Their numerical equality is demonstrated by the fact that all bodies fall to the earth with the same acceleration. Einstein expressed this equality in his Principle of Equivalence.

conditioned by the identity of inertia and weight, leads, in conjunction with the results of the special theory of relativity, to the conclusion that the laws governing the arrangement of solid bodies in space, when gravitational fields are present, do not correspond to the laws of Euclidean geometry. An analogous result follows for the motion of clocks. This brings us to the necessity for yet another generalization of the theory of space and time, because the direct interpretation of spatial and temporal coordinates by means of measurements obtainable with measuring rods and clocks now breaks down. That generalization of metric, which had already been accomplished in the sphere of pure mathematics through the researches of Gauss and Riemann, is essentially based on the fact that the metric of the special theory of relativity can still claim validity for small regions in the general case as well.

The process of development here sketched strips the space-time coordinates of all independent reality. The metrically real is now only given through the combination of the space-time coordinates with the mathematical quantities which describe the gravitational field.

There is yet another factor underlying the evolution of the general theory of relativity. As Ernst Mach insistently pointed out, the Newtonian theory is unsatisfactory in the following respect; if one considers motion from the purely descriptive, not from the causal, point of view, it only exists as relative motion of things with respect to one another. But the acceleration which figures in Newton's equations of motion is unintelligible if one starts with the concept of relative motion. It compelled Newton to invent a physical space in relation to which acceleration was supposed to exist. This introduction *ad hoc* of the concept of absolute space, while logically unexceptionable, nevertheless seems unsatisfactory. Hence Mach's attempt to alter the mechanical equations in such a way that the inertia of bodies is traced back to relative motion on their part not as against absolute space but as against the totality of other ponderable bodies. In the state of knowledge then existing, his attempt was bound to fail.

The posing of the problem seems, however, entirely reasonable. This line of argument imposes itself with considerably enhanced force in relation to the general theory of relativity, since, according to that theory, the physical properties of space are affected by ponderable matter. In my opinion the general theory of relativity can solve this problem satisfactorily only if it regards the world as spatially closed. The mathematical results of the theory force one to this view, if one believes that the mean density of ponderable matter in the world possesses some finite value, however small.

CHAPTER 22

THE QUANTUM THEORY

76. From MAX PLANCK'S Nobel Prize address, *The Origin and Development of the Quantum Theory* (1920)

When I recall the days of twenty years ago, when the conception of the physical quantum of 'action' was first beginning to disentangle itself from the surrounding mass of available experimental facts, and when I look back upon the long tortuous road which finally led to its disclosure, this development strikes me at times as a new illustration of Goethe's saying, that 'man errs, so long as he is striving'. And all the mental effort of an assiduous investigator must indeed appear vain and hopeless, if he does not occasionally run across striking facts which form incontrovertible proof of the truth he seeks, and show him that after all he has moved at least one step nearer to his objective. The pursuit of a goal, the brightness of which is undimmed by initial failure, is an indispensable condition, though by no means a guarantee, of final success.

In my own case such a goal has been for many years the solution of the question of the distribution of energy in the normal spectrum of radiant heat. The discovery by Gustav Kirchhoff that the quality of the heat radiation produced in an enclosure surrounded by any emitting or absorbing bodies whatsoever, all at the same temperature, is entirely independent of the nature of such bodies, established the existence of a universal function, which depends only upon the temperature and the wave-length, and is entirely independent of the particular properties of the substance. And the discovery of this remarkable function promised a deeper insight into the relation between energy and temperature, which is the principal problem of thermodynamics and therefore also of the entire field of molecular physics.

[Planck describes how he obtained a radiation formula containing two universal constants; the first was 'the value of the electrical elementary charge', i.e. the charge on the electron, and the second he now discusses.]

Much less simple than that of the first was the interpretation of the second universal constant of the radiation law, which, as the product of energy and time (amounting on a first calculation to $6.55.10^{-27}$ erg. sec.) I called the elementary quantum of action. While this constant was absolutely indispensable to the attainment of a correct expression for entropy—for only with its aid could be determined the magnitude of the 'elementary region' or 'range' of probability, necessary for the statistical treatment of the problem—it obstinately withstood all attempts at fitting it, in any suitable form, into the frame of the classical theory. So long as it could be regarded as infinitely small, that is to say for large values of energy or long periods of time, all went well; but in the general case a difficulty arose at some point or other, which became the more pronounced the weaker and the more rapid the oscillations. The failure of all attempts to bridge this gap soon placed one before the dilemma: either the quantum of action was only a fictitious magnitude, and, therefore, the entire deduction from the radiation law was illusory and a mere juggling with formulae, or there is at the bottom of this method of deriving the radiation law some true physical concept. If the latter were the case, the quantum would have to play a fundamental role in physics, heralding the advent of a new state of things, destined, perhaps, to transform completely our physical concepts which since the introduction of the infinitesimal calculus by Leibniz and Newton have been founded upon the assumption of the continuity of all causal chains of events.

Experience has decided for the second alternative. But that the decision should come so soon and so unhesitatingly was due not to the examination of the law of distribution of the energy of heat radiation, still less to my special deduction of this law, but to the steady progress of the work of those investigators who have applied the concept of the quantum of action to their researches.

77. From NIELS BOHR's article, *Discussion with Einstein on Epistemological Problems in Atomic Physics* (1949)

[At the Solvay conference in Brussels in October 1927, the interpretation of the quantum theory, devised by Bohr and his Copenhagen school, was subjected to the criticism of Einstein. The photon, which is a quantum of light, behaves both as a particle, detectable on a photographic plate, and as waves, detectable by an interference pattern. It therefore differs fundamentally from any object we experience in daily life. Bohr maintained in his theory of complementarity that, although a photon behaves simultaneously both as a particle and waves, we can observe it only in one of its aspects at one time. If we wish to observe it as a particle we must choose one type of apparatus; if as waves, another type.]

The extent to which renunciation of the visualization of atomic phenomena is imposed upon us by the impossibility of their subdivision is strikingly illustrated by the following example to which Einstein very early called attention and often has reverted. If a semi-reflecting mirror is placed in the way of a photon, leaving two possibilities for its direction of propagation, the photon may either be recorded on one, and only one, of two photographic plates situated at great distances in the two directions in question, or else we may, by replacing the plates by mirrors, observe effects exhibiting an interference between the two reflected wave-trains. In any attempt of a pictorial representation of the behaviour of the photon we would, thus, meet with the difficulty: to be obliged to say, on the one hand, that the photon always chooses *one* of the two ways and, on the other hand, that it behaves as if it had passed *both* ways.

It is just arguments of this kind which recall the impossibility of subdividing quantum phenomena and reveal the ambiguity in ascribing customary physical attributes to atomic objects....

These problems were instructively commented upon from different sides at the Solvay meeting, in the same session where Einstein raised his general objections. On that occasion an interesting discussion arose also about how to speak of the appearance of phenomena for which only predictions of statistical character can be made. The question was whether, as to the occurrence of individual effects, we should adopt a terminology proposed by Dirac, that we were concerned with a choice on the part of 'nature' or, as suggested by Heisenberg, we should say that we have to do with a choice on the part of the 'observer' constructing the measuring instruments and reading their recording. Any such terminology would, however, appear dubious since, on the one hand, it is hardly reasonable to endow nature with volition in the ordinary sense, while, on the other hand, it is certainly not possible for the observer to influence the events which may appear under the conditions he has arranged. To my mind, there is no other alternative than to admit that, in this field of experience, we are dealing with individual phenomena and that our possibilities of handling the measuring instruments allow us only to make a choice between the different complementary types of phenomena we want to study.

78. From WERNER HEISENBERG'S lecture, *Recent Changes in the Foundations of Exact Science* (1934)

[Heisenberg maintains that the theory of relativity and the quantum theory have effected permanent changes in the foundations of physical science.]

This immediately raises the more general question of the finality of the changes wrought by modern physics on the foundations of exact science. We have to discuss whether the scientist will once and for all have to renounce all thought of an objective time scale common to all observers, and of objective events in time and space independent of observations on them. Perhaps recent developments represent only a passing crisis. I tend to the opinion, for which there seems to be the strongest evidence, that this renunciation will be final. I would like to begin with an analogy to support this statement. Previous to the beginnings of science in antiquity, the world was conceived as a flat disc, and only the discovery of America and the first circumnavigation of the world destroyed this belief for all time. Of course nobody had ever seen the edge of the world-disc, but just the same this 'end of the world' acquired form and substance through the legends and imaginings of man. We all know the theme of the ever enquiring man who wants to travel to the end of the world. Then, the question of 'the end of the world' had a definite and clear meaning, but the voyages of discovery of Columbus and Magellan made that question meaningless and transformed the ideas linked to it into fairy tales for ever afterwards. For all that, mankind did not renounce the idea of 'the end of the world' as a result of having explored the whole surface of the world—even to-day there are some unexplored parts—but the voyages of Columbus and Magellan gave clear proofs of the necessity to make use of new lines of approach. In accepting the spherical shape of the earth the loss of the old concept was not felt to be a loss. Similarly modern physics has taught us to do without the concepts of an absolute scale of time and of objective events in space and time. The meaning of these two concepts had never been confirmed by direct experience either, at least not as completely as we had believed. They, too, formed a hypothetical 'end of the world'. It must be stressed that the world of ideas which is to be destroyed simultaneously with these concepts of classical physics is much less living than that destroyed by Columbus or Copernicus. Hence the transition to our concept of the universe, wrought by modern physics, is less decisive than that of the fifteenth and sixteenth centuries. The convincing power of the quantum theory is by no means based on the fact that we may have surveyed all methods of measuring the position and velocity of an electron and that we have been unsuccessful in every case in circumventing the uncertainty relations. But the experimental results of say Compton, Geiger, and Bothe are such clear proof of the necessity of making use of the new lines of thought introduced by quantum theory, that the loss of concepts of classical physics no longer appears a loss. The real strength of modern physics, then, rests in its new lines of thought. The hope that new experiments will yet lead us back to objective events in time and space, or to absolute time, are about as well founded as the hope of discovering the end of the world somewhere in the unexplored regions of the Antarctic. This analogy may be further extended; Columbus's discoveries were immaterial to the geography of the Mediterranean countries, and it would be quite wrong to claim that the voyages of discovery of the famous Genoese had made obsolete the positive geographical knowledge of the day. It is equally wrong to speak to-day of a revolution in physics. Modern physics has changed nothing in the great classical disciplines of, for instance, mechanics, optics, and heat. Only the conception of hitherto unexplored regions, formed prematurely from a knowledge of only certain parts of the world, has undergone a decisive transformation. This conception, however, is always decisive for the future course of research.

CHAPTER 23

COSMOGONY

79. From EDWIN HUBBLE's *The Observational Approach to Cosmology* (1937)

[Hubble expresses doubts whether the red-shifts in the spectra of the nebulae really indicate that the universe is expanding.]

When Slipher, in his great pioneering work, assembled the first considerable lists of red-shifts, the observations were necessarily restricted to the brighter, nearer nebulae. Consequently, the shifts were moderately small (less than 1 per cent), and they were accepted without question as the familiar velocity-shifts. Attempts were immediately made to study the motions of the nebulae by the same methods used in the study of stellar motions. But later, after the 'velocity-distance relation' had been formulated, and Humason's observations of faint nebulae began to accumulate, the earlier, complete certainty of the interpretation began to fade.

The disturbing features were the facts that the 'velocities' reached enormous values and were precisely correlated with distance. Each million light-years of distance added a hundred miles per second to the 'velocity'. As Humason swept farther and farther out into space he reported 'velocities' of 5,000 miles per second, then 10,000 then 15,000. Finally, near the absolute limit of his spectrograph he recorded red-shifts of 13 and 14 per cent, 'velocities' of about 25,000 miles per second—around the earth in a second, out to the moon in 10 seconds, out to the sun in just over an hour. Red-shifts continue to increase beyond the range of the spectrograph, and, for the faintest nebulae that can be photographed, they are presumably about double the largest recorded shifts—the 'velocities' are about 50,000 miles per second. These quantities we are asked to accept as measuring a general recession of the nebulae, an expansion of the universe itself. The law of red-shifts then reads: the nebulae are receding from the earth, in all directions, with velocities that are proportional to their distances from the earth.

Well, perhaps the nebulae are all receding in this peculiar manner. But the notion is rather startling. The cautious observer naturally examines other possibilities before accepting the proposition even as a working hypothesis. He recalls the alternative formulation of the law of red-shifts—light loses energy in proportion to the distance it travels through space. The law, in this form, sounds quite plausible. Internebular space, we believe, cannot be entirely empty. There must be a gravitational field through which the light-quanta travel for many millions of years before they reach the observer, and there may be some interaction between the quanta and the surrounding medium. The problem invites speculation, and, indeed, has been carefully examined. But no satisfactory, detailed solution has been found. The known reactions have been examined, one after the other—and they have failed to account for the observations. Light *may* lose energy during its journey through space, but if so, we do not yet know how the loss can be explained.

The observer seems to face a dilemma. The familiar interpretation of red-shifts leads to rather startling conclusions. These conclusions can be avoided by an assumption which sounds plausible but which finds no place in our present body of knowledge. The situation can be described as follows. Red-shifts are produced either in the nebulae, where the light originates, or in the intervening space through which the light travels. If the source is in the nebulae, then red-shifts are probably velocity-shifts and the nebulae are receding. If the source lies in the intervening space, the explanation of red-shifts is unknown but the nebulae are sensibly stationary.

80. From ALBERT EINSTEIN'S *The Meaning of Relativity* (1950)

[Einstein first refers to the cosmological term, representing a force of repulsion, which he was obliged to insert into the field equations of general relativity to account for a static universe. He then explains why the red-shift must be taken as an indication that the universe is expanding.]

If Hubble's expansion had been discovered at the time of the creation of the general theory of relativity, the cosmologic member would never have been introduced. It seems now so much less justified to introduce such a member into the field equations, since its introduction loses its sole original justification—that of leading to a natural solution of the cosmologic problem. . . .

Some try to explain Hubble's shift of spectral lines by means other than the Doppler effect. There is, however, no support for such a conception in the known physical facts. According to such a hypothesis it would be possible to connect two stars, S_1 and S_2, by a rigid rod. Monochromatic light which is sent from S_1 to S_2 and reflected back to S_1 could arrive with a different frequency (measured by a clock on S_1) if the number of wave lengths of light along the rod should change with time on the way. This would mean that the locally measured velocity of light would depend on time, which would contradict even the special theory of relativity. Further it should be noted that a light signal going to and fro between S_1 and S_2 would constitute a 'clock' which would not be in a constant relation with a clock (e.g. an atomistic clock) in S_1. This would mean that there would exist no metric in the sense of relativity. This not only involves the loss of compre-hension of all those relations which relativity has yielded, but it also fails to concur with the fact that certain atomistic forms are not related by 'similarity' but by 'congruence' (the existence of sharp spectral lines, volumes of atoms, etc.). . . .

Hence one cannot but consider Hubble's discovery as an expansion of the system of stars.

81. From HERMANN BONDI'S *Cosmology* (1952)

[Bondi states the perfect cosmological principle and the theory of continuous creation based upon it.]

For in any theory which contemplates a changing universe, explicit and implicit assumptions must be made about the inter-actions between distant matter and local physical laws. These assumptions are necessarily of a highly arbitrary nature, and progress on such a basis can only be indefinite and uncertain. It may, however, be questioned whether such speculation is required. If the uniformity of the universe is sufficiently great none of these difficulties arises. The assumption that this is so is known as the perfect cosmological principle. It was introduced by Bondi and Gold (1948) in the form of the statement, that, apart from local irregularities the universe presents the same aspect from any place at any time. It was shown by Bondi and Gold that this single principle forms a sufficient basis for developing without ambiguity a cosmological theory capable of making definite and far-reaching physical statements agreeing with observation. . . .

The fundamental assumption of the theory is that the universe presents on the large scale an unchanging aspect. Since the universe must (on thermodynamic grounds) be expanding, new matter must be continually created in order to keep the density constant. As ageing nebulae drift apart, due to the general motion of expansion, new nebulae are formed in the intergalactic spaces by condensation of newly created matter. Nebulae of all ages hence exist with a certain frequency distribution. Astrophysical estimates of the age of our galaxy do not put it into a very rare class of nebulae.

The theory is deductive in the sense that its conclusions are derived from the cosmological principle, but the very powerful formulation of the principle employed dispenses with the need for additional assumptions.

A different approach has been proposed by Hoyle and will be discussed later in this chapter. In that formulation a suitable modification of the field equations of general relativity is taken as starting-point, so that the conclusions reached are very similar to those of the steady-state theory.

The importance to the theory of the powerful formulation of the cosmological principle makes it highly desirable to examine in detail the arguments for the acceptance of this principle. This examination reveals that the arguments supporting the usual narrow cosmological principle imply the validity of the wider perfect cosmological principle, according to which the large-scale aspect of the universe should not only be independent of the *position* of the observer but also of the *time* of making the observation. . . .

There is little doubt that the continual creation of matter necessary in this theory is the most revolutionary change proposed by it. There is, however, no observational evidence whatever contradicting continual creation at the rate demanded by the perfect cosmological principle. It is easily seen that this is, on the average,

$$3 \times (\text{mean density of matter in the universe})$$
$$\times \text{Hubble's constant} = 10^{-43} \text{ g./cm.}^3 \text{ sec.}$$

approximately. In other words, on an average the mass of a hydrogen atom is created in each litre of volume every 10^9 years. As will be seen later there are strong arguments showing that the creation rate does not vary widely between different places, so that the average rate given above has universal significance. It is clear that it is utterly impossible to observe directly such a rate of creation. There is therefore no contradiction whatever with the observations, an extreme extrapolation from which forms the principle of conservation of matter. The argument may be stated in the following terms: When observations indicated that matter was at least very nearly conserved it seemed simplest (and therefore most scientific) to assume that the conservation was absolute. But when a wider field is surveyed then it is seen that this apparently simple assumption leads to the great complications discussed in connexion with the formulation of the perfect cosmological principle. The principle resulting in greatest overall simplicity is then seen to be not the principle of conservation of matter but the perfect cosmological principle with its consequence of continual creation. From this point of view continual creation is the simplest and hence the most scientific extrapolation from the observations.

CHAPTER 24

THE TWENTIETH CENTURY

82. From ISAAC NEWTON'S *Opticks* (1704)

[This passage, in which Newton states his belief that the ultimate physical reality consists of massy atoms, is taken from the last Query in his *Opticks*.]

All these things being consider'd, it seems probable to me, that God in the Beginning form'd Matter in solid, massy, hard, impenetrable, movable Particles, of such Sizes and Figures, and with such other Properties, and in such Proportion to Space, as most conduced to the End for which he form'd them; and that these primitive Particles being Solids, are incomparably harder than any porous Bodies compounded of them; even so very hard, as never to wear or break in pieces; no ordinary Power being able to divide what God himself made one in the first Creation. While the Particles continue entire, they may compose Bodies of one and the same Nature and Texture in all Ages: But should they wear away, or break in pieces, the Nature of Things depending on them, would be changed. Water and Earth, composed of old worn Particles and Fragments of Particles, would not be of the same Nature and Texture now, with Water and Earth composed of entire Particles in the Beginning. And therefore, that Nature may be lasting, the Changes of corporeal Things are to be placed only in the various Separations and new Associations and Motions of these permanent Particles; compound Bodies being apt to break, not in the midst of solid Particles, but where those Particles are laid together, and only touch in a few points.

83. From ERNST MACH's lecture, *The Economical Nature of Physical Enquiry* (1882)

[Mach states his view that science is concise description and that atoms are merely imaginary tools for this purpose.]

The communication of scientific knowledge always involves description, that is, a mimetic reproduction of facts in thought, the object of which is to replace and save the trouble of new experience. Again, to save the labour of instruction and of acquisition, concise, abridged description is sought. This is really all that natural laws are. Knowing the value of the acceleration of gravity, and Galileo's laws of descent, we possess simple and compendious directions for reproducing in thought all possible motions of falling bodies. A formula of this kind is a complete substitute for a full table of motions of descent, because by means of the formula the data of such a table can be easily constructed at a moment's notice without the least burdening of the memory. . . .

When we look over a province of facts for the first time, it appears to us diversified, irregular, confused, full of contradictions. We first succeed in grasping only single facts, unrelated with the others. The province, as we are wont to say, is not *clear*. By and by we discover the simple, permanent elements of the mosaic, out of which we can mentally construct the whole province. When we have reached a point where we can discover everywhere the same facts, we no longer feel lost in this province; we comprehend it without effort; it is *explained* for us. . . .

Those elements of an event which we call 'cause and effect' are certain salient features of it, which are important for its mental reproduction. Their importance wanes and the attention is transferred to fresh characters the moment the event or experience in question becomes familiar. If the connexion of such features strikes us as a necessary one, it is simply because the interpolation of certain intermediate links with which we are very familiar, and which possess, therefore, higher authority for us, is often attended with success in our explanations. That *ready* experience fixed in the mosaic of the mind with which we meet new events, Kant calls an innate concept of the understanding (*Verstandesbegriff*).

The grandest principles of physics, resolved into their elements, differ in no wise from the descriptive principles of the natural historian. . . .

When a geometer wishes to understand the form of a curve, he first resolves it into small rectilinear elements. In doing this, however, he is fully aware that these elements are only provisional and arbitrary devices for comprehending in parts what he cannot comprehend as a whole. When the law of the curve is found he no longer thinks of the elements. Similarly, it would not become physical science to see in its self-created, changeable, economical tools, molecules and atoms, realities behind phenomena, forgetful of the lately acquired sapience of her older sister, philosophy, in substituting a mechanical mythology for the old animistic or metaphysical scheme, and thus creating no end of suppositious problems. The atom must remain a tool for representing phenomena, like the functions of mathematics. Gradually, however, as the intellect, by contact with its subject-matter, grows in discipline, physical science will give up its mosaic play with stones and will seek out the boundaries and forms of the bed in which the living stream of phenomena flows. The goal which it has set itself is the *simplest* and *most economical* abstract expression of facts.

84. From HENRI POINCARÉ's *Science and Hypothesis* (1902)

[Poincaré maintains that scientific theories merely express relations between phenomena and hence that, although superseded by others with different imagery, they may still retain their usefulness.]

The ephemeral nature of scientific theories takes by surprise the man of the world. Their brief period of prosperity ended, he sees them abandoned one after another; he sees ruins piled upon ruins; he predicts that the theories in fashion to-day will in a short time succumb in their turn, and he concludes that they are absolutely in vain. This is what he calls the *bankruptcy of science*.

His scepticism is superficial; he does not take into account the object of scientific theories and the part they play, or he would understand that the ruins may be still good for something. No theory seemed established on firmer ground than Fresnel's, which attributed light to the movements of the ether. Then if Maxwell's theory is to-day preferred, does that mean that Fresnel's work was in vain? No; for Fresnel's object was not to know whether there really is an ether, if it is or is not formed of atoms, if these atoms really move in this way or that; his object was to predict optical phenomena.

This Fresnel's theory enables us to do to-day as well as it did before Maxwell's time. The differential equations are always true, they may be always integrated by the same methods, and the results of this integration still preserve their value. It cannot be said that this is reducing physical theories to simple practical recipes; these equations express relations, and if the equations remain true, it is because the relations preserve their reality. They teach us now, as they did then, that there is such and such a relation between this thing and that; only, the something which we then called *motion*, we now call *electric current*. But these are merely names of the images we substituted for the real objects which Nature will hide for ever from our eyes. The true relations between these real objects are the only reality we can attain, and the sole condition is that the same relations shall exist between these objects as between the images we are forced to put in their place. If the relations are known to us, what does it matter if we think it convenient to replace one image by another?

85. From ALBERT EINSTEIN's lecture, *On the Method of Theoretical Physics* (1933)

[Einstein maintains that physical theories are free inventions of the mind and not unique inductions from the facts.]

Let us now cast an eye over the development of the theoretical system, paying special attention to the relations between the content of the theory and the totality of empirical fact. We are concerned with the eternal antithesis between the two inseparable components of our knowledge, the empirical and the rational, in our department.

We reverence ancient Greece as the cradle of western science. Here for the first time the world witnessed the miracle of a logical system which proceeded from step to step with such precision that every single one of its propositions was absolutely indubitable—I refer to Euclid's geometry. This admirable triumph of reasoning gave the human intellect the necessary confidence in itself for its subsequent achievements. If Euclid failed to kindle your youthful enthusiasm, then you were not born to be a scientific thinker.

But before mankind could be ripe for a science which takes in the whole of reality, a second fundamental truth was needed, which only became common property among philosophers with the advent of Kepler and Galileo. Pure logical thinking cannot yield us any knowledge of the empirical world; all knowledge of reality starts from experience and ends in it. Propositions arrived at by purely logical means are completely empty as regards reality. Because Galileo saw this, and particularly because he drummed it into the scientific world, he is the father of modern physics—indeed, of modern science altogether.

If then, experience is the alpha and the omega of all our knowledge of reality, what is the function of pure reason in science?

A complete system of theoretical physics is made up of concepts, fundamental laws which are supposed to be valid for those concepts and conclusions to be reached by logical deduction. It is these conclusions which must correspond with our separate experiences; in any theoretical treatise their logical deduction occupies almost the whole book.

This is exactly what happens in Euclid's geometry, except that there the fundamental laws are called axioms and there is no question of the conclusions having to correspond to any sort of experience. If, however, one regards Euclidean geometry as the science of the possible mutual relations of practically rigid bodies in space, that is to say, treats it as a physical science, without abstracting from its original empirical content, the logical homogeneity of geometry and theoretical physics becomes complete.

We have thus assigned to pure reason and experience their places in a theoretical system of physics. The structure of the system is the work of reason; the empirical contents and their mutual relations must find their representation in the conclusions of the theory. In the possibility of such a representation lie the sole value and justification of the whole system, and especially of the concepts and fundamental principles which underlie it. Apart from that, these latter are free inventions of the human intellect, which cannot be justified either by the nature of that intellect or in any other fashion *a priori*.

These fundamental concepts and postulates, which cannot be further reduced logically, form the essential part of a theory, which reason cannot touch. It is the grand object of all theory to make these irreducible elements as simple and as few in number as possible, without having to renounce the adequate representation of any empirical content whatever.

The view I have just outlined of the purely fictitious character of the fundamentals of scientfic theory was by no means the prevailing one in the eighteenth and nineteenth centuries. But it is steadily gaining ground from the fact that the distance in thought between the fundamental concepts and laws on one side and, on the other, the conclusions which have to be brought into relation with our experience grows larger and larger, the simpler the logical structure becomes—that is to say, the smaller the number of logically independent conceptual elements which are found necessary to support the structure.

Newton, the first creator of a comprehensive, workable system of theoretical physics, still believed that the basic concepts and laws of his system could be derived from experience. This is no doubt the meaning of his saying, *hypotheses non fingo*.

Actually the concepts of time and space appeared at that time to present no difficulties. The concepts of mass, inertia, and force, and the laws connecting them, seemed to be drawn directly from experience. Once this basis is accepted, the expression for the force of gravitation appears derivable from experience, and it was reasonable to expect the same in regard to other forces.

We can indeed see from Newton's formulation of it that the concept of absolute space, which comprised that of absolute rest, made him feel uncomfortable; he realized that there seemed to be nothing in experience corresponding to this last concept. He was also not quite comfortable about the introduction of forces operating at a distance. But the tremendous practical success of his doctrines may well have prevented him and the physicists of the eighteenth and nineteenth centuries from recognizing the fictitious character of the foundations of his system.

The natural philosophers of those days were, on the contrary, most of them possessed with the idea that the fundamental concepts and postulates of physics were not in the logical sense free inventions of the human mind but could be deduced from experience by 'abstraction'—that is to say, by logical means. A clear recognition of the erroneousness of this notion really only came with the general theory of relativity, which showed that one could take account of a wider range of empirical facts, and that, too, in a more satisfactory and complete manner, on a foundation quite different from the Newtonian. But quite apart from the question of the superiority of one or the other, the fictitious character of fundamental principles is perfectly evident from the fact that we can point to two essentially different principles, both of which correspond with experience to a large extent; this proves at the same time that every attempt at a logical deduction of the basic concepts and postulates of mechanics from elementary experiences is doomed to failure.

If, then, it is true that the axiomatic basis of theoretical physics cannot be extracted from experience but must be freely invented, can we ever hope to find the right way? Nay, more, has this right way any existence outside our illusions? Can we hope to be guided safely by experience at all when there exist theories (such as classical mechanics) which to a large extent do justice to experience, without getting to the root of the matter? I answer without hesitation that there is, in my opinion, a right way, and that we are capable of finding it. Our experience hitherto justifies us in believing that nature is the realization of the simplest conceivable mathematical ideas. I am convinced that we can discover by means of purely mathematical constructions the concepts and the laws connecting them with each other, which furnish the key to the understanding of natural phenomena. Experience may suggest the appropriate mathematical concepts, but they most certainly cannot be deduced from it. Experience remains, of course, the sole criterion of the physical utility of a mathematical construction. But the creative principle resides in mathematics. In a certain sense, therefore, I hold it true that pure thought can grasp reality, as the ancients dreamed.

86. From ARTHUR EDDINGTON'S *The Philosophy of Physical Science* (1938)

[Eddington's main thesis is that 'all the laws of nature that are usually classed as fundamental can be foreseen wholly from epistemological considerations. They correspond to *a priori* knowledge, and are therefore wholly subjective.' In the following passage he banteringly suggests that most physicists pay lip service to this idea but do not really believe it.]

I am about to turn from the scientific to the philosophical setting of scientific epistemology. This is accordingly a suitable place at which to make a comparison with the most commonly accepted view of scientific philosophy. The following statement is fairly typical:

That science is concerned with the rational correlation of experience rather than with the discovery of fragments of absolute truth about an external world is a view which is now widely accepted.[1]

I think that the average physicist, in so far as he holds any philosophical view at all about his science, would assent. The phrase 'rational correlation of experience' has a savour of orthodoxy which makes it a safe gambit for applause. The repudiation of more adventurous aims gives a comfortable feeling of modesty—all the more agreeable if we fancy that someone else is being told off. For my own part I accept the statement, provided that 'science' is understood to mean 'physics'. It has taken me nearly twenty years to accept it; but by steady mastication during that period I have managed to swallow it all down bit by bit. Consequently I am rather flabbergasted by the light-hearted way in which this pronouncement, carrying the most profound implications both for philosophy and for physics, is commonly made and accepted.

I have no serious quarrel with the average physicist over his philosophical creed—except that he forgets all about it in practice. My puzzle is why a belief that physics is concerned with the correlation of experience and not with absolute truth about the external world should usually be accompanied by a steady refusal to treat theoretical physics as a description of correlations of experience and an insistence on treating it as a description of the contents of an absolute objective world. If I am in any way heterodox, it is because it seems to me a consequence of accepting the belief, that we shall get nearer to whatever truth is to be found in physics by seeking and employing conceptions suitable for the

[1] Unsigned review, *Phil. Mag.* 25 (1938), 814.

expression of correlations of experience instead of conceptions suitable for the description of an absolute world.

The statement evidently means that the methods of physics are incapable of discovering fragments of absolute truth about an external world; for we should have no right to withold from mankind the absolute truth about the external world if it were within our reach. If the laboratories, built and endowed at great expense, could assist in the discovery of absolute truth about the external world, it would be reprehensible to discourage their use for this purpose. But the assertion that the methods of physics cannot reveal absolute (objective) truth or even fragments of absolute truth, concedes my main point that the knowledge obtained by them is wholly subjective. Indeed it concedes it far too readily; for the assertion is one that ought only to be made after prolonged investigation. As I have pointed out, sciences other than physics and chemistry are not so limited in their scope. The discovery of unmistakeable signs of intelligent life on another planet would be hailed as an epoch-making astronomical achievement; it can scarcely be denied that it would be the discovery of a fragment of absolute truth about the world external to us.

Keeping to physics, the commonly accepted scientific philosophy is that it is not concerned with the discovery of absolute truth about the external world, and its laws are not fragments of absolute truth about the external world, or, as I have put it, they are not laws of the objective world. What then are they, and how is it that we find them in our correlations of experience? Until we can see, by an examination of the procedure of correlation of our observational experience, how these highly complex laws have got into it subjectively, it seems premature to accept a philosophy which cuts us off from all other possible explanations of their origin. This is the examination that we have been conducting.

The end of our journey is rather a bathos after so much toil. Instead of struggling up to a lonely peak, we have reached an encampment of believers, who tell us 'That is what we have been asserting for years'. Presumably they will welcome with open arms the toilworn travellers who have at last found a resting place in the true faith. All the same I am a bit dubious about that welcome. Perhaps the assertion, like many a religious creed, was intended only to be recited and applauded. Anyone who *believes* it is a bit of a heretic.

87. From RAPHAEL DEMOS' article, *Doubts about Empiricism* (1947)

[This is part of a light-hearted, but penetrating, attack on current scientific philosophies which are based on empiricism, i.e. the belief that all knowledge must be derived from sense data.]

My beliefs during the first stage of my philosophical career were a mixed brew of ingredients taken from the Greek and Christian traditions. My tastes were conservative and even reactionary. I believed in the reality of substance, material and mental; I held that there are universal and necessary connections in nature which can be known. In short, I was a naïve objectivist about things and about structures. I was a realist about values too. I believed that there are such traits in nature as good and bad, right and wrong, beauty and ugliness, independently of my preferences. I was convinced further that goodness was a supreme causal agent; that—to paraphrase the familiar quotation—righteousness is power. I believed in God. With Plato (in the *Phaedo*) I maintained that causality is not only efficient but final too; that nature exhibits both a mechanical and a moral order. And these two propositions were, to my view, but twin aspects of the one proposition that nature will not deceive my expectations.

How did I come to believe all this? My convictions came out of a curious amalgam of reason, experience and faith. By experience I meant not only sense-perception but also what has been called religious experience; I also meant feeling as furnishing an acquaintance with values. By reason I meant a power which apprehends principles and laws, natural and moral. By faith I meant a faculty which provides me with my premises. Thus it was by faith primarily that I obtained my belief in God as a moral and personal power within nature and transcending it.

All this and more I cherished naïvely until I was shocked out of it by those of my friends who were philosophers of the scientific camp. They told me that I must believe nothing which did not conform with the canons of scientific method. Above all, I must have no faith in faith. Science is the determined enemy of all authority, tradition and dogmatism. I must submit my beliefs to the test of rational evidence, that is to say, to the test of the senses. I realized that neither religious 'experience' nor feelings are evidence. Thus I gave up faith altogether and modified what

I called experience and reason. Experience now was strictly identified with sensation; and reason meant either the comparison of hypotheses with the data of sense or the analysis of concepts. As my windows to the world became fewer or smaller or otherwise changed in shape, my world too became different. I decided that substance does not exist since it is not a datum of sense. I had secret misgivings of course. The view appeared to me incredible that there were qualia without things which they qualified, that there were events in which nothing materialized. It all seemed like the episode in Peter Pan's Neverland, when, after the Redskins destroyed the nest, the eggs still remained suspended in mid-air. But I swept these misgivings aside as obviously weak-minded and bourgeois. My belief in God was clearly a superstition and had to go the way of beliefs in ghosts and fairies. Values, too, were out; they were nothing more than projections of feelings, just as my religious beliefs were rationalizations of desire. I had been subjectivistic as well as dogmatic and reactionary. Certainly my sins had been many-sided and now I was truly remorseful. . . .

And now I must make a humiliating confession. I could not really believe what I professed to believe. While my good, scientific self envisaged sense-qualia and their configurations, my bad, practical self unconsciously went on behaving as though there were material objects, other minds, necessary connections and the rest. For instance, in the summer I would still make arrangements to prepare against the cold in the winter (by buying coal) even though I knew that I did not know of any necessary connection between past and future. While I believed human beings to be nothing more than collections of material particles, essentially not different from stones, I went on treating my fellow-humans with respect as though they were souls endowed with infinite worth. I realized that I was becoming a divided personality, and a schizophrenic. I became afraid for myself, having recalled the depression to which Hume had been plunged by his empiricism. So I called on a psychoanalyst and laid my troubles before him. When, following the usual technique, he suggested that I start trains of free association with the ideas that bothered me, I babbled: 'matter, madder, hatter' and similar foolishness which I will spare the reader. The doctor's diagnosis of my troubles was all too soothing. He explained that modern empiricism and positivism were but echoes of the Puritan temper in the field of thought. Puritanism, of course, is the great disease of the psyche. Just as religious puritans starved their passions, so had I starved my animal beliefs. Empiricism, he said, is the asceticism of the intellect. He asked me to compare my earlier pagan world, populated with full-blooded substances, to my present austere world of threadbare and anemic sense-data. Empiricism had transformed my flaming garden into a desert. My doctor went so far as to compare me in my present reduced state of belief to one of those victims of bombed-out areas who could carry all their goods in a satchel. The way back to health was obvious: let me overthrow my empirical censor and indulge my natural propensities for belief to the full.

What comforting and yet what poisonous advice! The doctor was actually proposing that I should let my desires guide my beliefs! No, no, I cried, rushing out of his office; better suffer with doubt than prosper with faith. Like John Bunyan I now felt with relief that a great burden had fallen from my shoulders. I was free, free from fear and superstition. To paraphrase one of my masters: there were no longer any spirits, essences, transcendental ideas, immaterial angels and principles to haunt my life with their terrors. The important thing was security. Condemned as I was to inhabit a waste land, at least I knew I was safe; where there is nothing, nothing can hurt me.

Suddenly, I had a bright idea. What had caused me—I asked myself—to reject my earlier common sense and so to be mired in doubt? Why, the conception of scientific method and the generalized doctrine of empiricism. In short, I had been plunged into the darkness of doubt by the glare of certainty. Here, then, is a shaft of light, I said to myself; let me turn to its source for life and heat. Imagine my disappointment when I discovered that the source of the light was not the sun, not even the other side of the moon, but some smoky sickly fire concocted by witches. The doctrines I just cited were nothing more than uncriticized beliefs. My philosophical friends from the naturalistic camp had taken over without questioning the conviction that science is knowledge and that it is the only knowledge; and that sense-perception is our only contact with reality. When I heard them urging everybody to think in philosophy as the scientist does in his field, I smelled an appeal to a new and a more terrible authority. . . .

I had been taught that, unlike religion, science takes nothing on faith. I accepted that statement on faith, but when I examined scientific procedure I was disillusioned. Science establishes its generalizations by an appeal to the inductive principle.[1] Now, what is the evidence for the latter? Not experience surely—Hume's word is final on that point. Scientists check all theories by experience, save the one overarching theory of the uniformity of nature. I was struck with the fact that my naturalist friends proclaimed themselves empiricists when they could offer no empirical evidence for their basic premise. I was even more struck with the fact that while they proclaimed themselves opponents of all uncriticized belief, they had failed to submit their own theory of knowledge to a critical scrutiny. When one of them happened to defend commitment to the inductive principle as a sort of desperate gamble, I was vaguely reminded of Pascal's wager as an argument for the existence of God. Naturalists condescendingly suggest that people resort to the idea of God because it relieves them from anxiety. In the same spirit, it might be suggested that people resort to the principle of the uniformity of nature because it makes them feel at home in the universe, by picturing the latter as tidy and well-behaved. Should I be told that the inductive principle is validated pragmatically because it has worked so far, I would retort that so to argue is to beg the question. What I want to know is whether the principle will work in the future; and, unless I assume the inductive principle, the fact that it has worked in the past goes for nothing. . . .

From Hume down (not to mention Sextus Empiricus) philosophers have made the stupendous assumption that only what is given *to sense* is given. My point at the moment is not that they are wrong, or even that their arguments are unsound; it is that they give no arguments at all. They simply use the words experience and sense-experience as equivalent, without indicating that a definite transition has been made. Surely when the point is so far-reaching in its consequences, it should have been supported by reasons. Of course, all conscious processes are experiences in the initial sense that they are data of conscious awareness. But that is not what is in question. The point rather is that among all the conscious experiences, only sensations are directly cognitive, or at any rate, furnish data for knowledge. Here are sensations with their qualia, feelings with their values,

[1] That it is possible to obtain a valid generalization from particular cases.

visions of God, rational insights: on what basis have these philosophers excluded (as subjective) every experience but that of sense? And yet Plato and others had held sense to be illusory; there was a real issue to be met....

It has been urged in defence of empiricism that sensa satisfy the criterion of intersubjectivity, as other conscious experiences (like feelings and religious insights) do not. But on what grounds is intersubjectivity set up as a criterion? Surely it is reasonable to expect that when somebody like an artist has a unique perspective, his data too would be unique to himself, and still be veridical data. Intersubjectivity (along with other criteria such as simplicity, measurability and predictive potency) is simply an arbitrary rule to control the cognitive behaviour of the scientists; it is part of their *mores*. The tribes of the South Pacific practise polygamy; some nations practise communism; and the scientists have their own peculiar rules to which they conform in their activity of believing....

My investigation of positivism was more brief because the doctrine itself is still in flux. First, I asked myself what proof was offered for the proposition that all meaning is empirical. None that I could find; none had even been attempted so far as I could see. Surely no empirical justification for the positivist doctrine could be found in the traditional-historical use of the term meaning; in fact, many philosophers had used the term to mean non-empirical meaning. Otherwise, what was all the fuss about? Did the positivists maintain that only when meaning is empirical can it be clear and exact? Granted; then the positivists lay down a demand for clarity and exactness. And if so, they are simply pressing upon us their arbitrary preferences and valuations; they are poets, and their place is in the empyrean, not down here in the valleys of prosaic philosophizing. Similarly, my operationalist friends insisted that religion *ought* to follow the example of science and to restrict itself to terms whose meaning is operational. Again, I was unable to fathom the operational meaning of the categorical imperatives proclaimed by my friends....

With this I concluded my review of my new beliefs; the upshot was that I had been made to move out of one church in order to be forced into another. Just as Luther, after denying the authority of the Pope, ended up by affirming the authority of the Bible, my new friends set up their own dogmas in place of those of religion.

Science was the new faith, and its credo could be stated as follows:

1. I believe in the principle of induction, omnipresent and omnipotent.

2. I believe in the uniqueness of sensation as an insight into the real.

3. I believe in Occam's razor as the mediating principle between theories and facts.

Scholium: These three principles are indivisible; three in one, one in three.

4. I believe in the intersubjective communion of the scientists-saints.

5. I believe in the coming of the Kingdom of Knowledge when all events will be systematically explained.

6. I believe in the resurrection of time from the tomb of the past, in a body which is spiritual because it is a construct of the mind.

7. I believe the above articles to be the fixed and final faith forever. Failure to accept them is a failure of nerve.

88. From Vladimir Ilyich Lenin's *Materialism and Empirio-Criticism* (1908)

[Lenin upholds materialism and pours scorn on the idealism of Kant and Mach.]

In his *Ludwig Feuerbach*, Engels declares that the fundamental philosophical trends are materialism and idealism. Materialism regards nature as primary and spirit as secondary; it places being first and thought second. Idealism holds the contrary view. This root distinction between the 'two great camps' into which the philosophers of the 'various schools' of idealism and materialism are divided Engels takes as the cornerstone, and he directly charges with 'confusion' those who use the terms idealism and materialism in any other way....

Engels continues:

'The most telling refutation of this [idealism] as of all other philosophical fancies is practice, viz., experiment and industry. If we are able to prove the correctness of our conception of a natural process by making it ourselves, bringing it into being out of its conditions and using it for our own purposes into the bargain, then there is an end of the Kantian incomprehensible.... The chemical substances produced in the bodies of plants and animals remained just such "things-in-themselves" until organic chemistry began to produce them one after another, whereupon the "thing-in-itself" became a thing for us, as, for instance, alizarin, the colouring matter of the madder, which we no longer trouble to grow in the madder roots in the field, but produce more cheaply and simply from coal tar....'

What is the kernel of Engels' objections? Yesterday we did not know that coal tar contained alizarin. Today we learned that it does. The question is, did coal tar contain alizarin yesterday?

Of course it did. To doubt it would be to make a mockery of modern science.

And if that is so, three important epistemological conclusions follow:

(1) Things exist independently of our consciousness, independently of our perceptions, outside of us, for it is beyond doubt that alizarin existed in coal tar yesterday and it is equally beyond doubt that yesterday we knew nothing of the existence of this alizarin and received no sensations from it.

(2) There is definitely no difference in principle between the phenomenon and the thing-in-itself, and there can be no such difference. The only difference is between what is known and what is not yet known. And philosophical inventions of specific boundaries between the one and the other, inventions to the effect that the thing-in-itself is 'beyond' phenomena (Kant), or that we can or must fence ourselves off by some philosophical partition from the problem of a world which in one part or another is still unknown but which exists outside us (Hume)—all this is the sheerest nonsense, *Schrulle*, evasion, invention.

(3) In the theory of knowledge, as in every other branch of science, we must think dialectically, that is, we must not regard our knowledge as ready-made and unalterable, but must determine how *knowledge* emerges from ignorance, how incomplete, inexact knowledge becomes more complete and more exact.

Once we accept the point of view that human knowledge develops from ignorance, we shall find millions of examples of it just as simple as the discovery of alizarin in coal tar, millions of observations not only in the history of science and technology but in the everyday life of each and every one of us that illustrate the transformation of 'things-in themselves' into 'things-for-us', the appearance of 'phenomena' when our sense-organs experience a jolt from external objects, the disappearance of 'phenomena' when some obstacle prevents the action upon our sense-organs of an object which we know to exist. The sole and unavoidable deduction to be made from this—a deduction which all of us make in everyday practice and which materialism deliberately places at the foundation of its epistemology—is that outside us, and independently of us, there exist objects, things and bodies and that our perceptions are images of the external world. Mach's converse theory (that bodies are complexes of sensations) is nothing but pitiful idealist nonsense.

89. From FREDERICK ENGELS' *Dialectics of Nature* (written *c.* 1872–82, first published 1927)

[Engels states the laws of materialist dialectics.]

It is, therefore, from the history of nature and human society that the laws of dialectics are abstracted. For they are nothing but the most general laws of these two aspects of historical development, as well as of thought itself. And indeed they can be reduced in the main to three:

The law of the transformation of quantity into quality and *vice versa*;

The law of the interpenetration of opposites;

The law of the negation of the negation.

All three are developed by Hegel in his idealist fashion as mere laws of *thought*: the first, in the first part of his *Logic*, in the *Doctrine of Being*; the second fills the whole of the second and by far the most important part of his *Logic*, the *Doctrine of Essence*; finally the third figures as the fundamental law for the construction of the whole system. The mistake lies in the fact that these laws are foisted on nature and history as laws of thought, and not deduced from them. This is the source of the whole forced and often outrageous treatment; the universe, willy-nilly, is made out to be arranged in accordance with a system of thought which itself is only the produce of a definite stage of evolution of human thought. If we turn the thing round, then everything becomes simple, and the dialectical laws that look so extremely mysterious in idealist philosophy at once become simple and clear as noonday.

90. Letter by M. POLANYI to *Nature* on *The Cultural Significance of Science* (1940)

[The letter represents one of the many shots fired in the battle between those who advocate the planning of pure science by the State and those who oppose this.]

The admonition in *Nature* of December 28 to abandon 'once and for all the belief that science is set apart from all other social interests as if it possessed a peculiar holiness' expresses very precisely the opposite of what many men of science consider to be their duty. I, for one, can recognize nothing more holy than scientific truth, and consider it a danger to science and to humanity if the pursuit of pure science, regardless of society, is denied by a representative organ of science.

For the last ten years we have been presented by an influential school of thought with phrases about the desirability of a social control of science, accompanied by attacks on the alleged snobbishness and irresponsibility of scientific detachment. The 'social control of science' has proved a meaningless phrase. Science exists only to that extent to which the search for truth is not socially controlled. And therein lies the purpose of scientific detachment. It is of the same character as the independence of the witness, of the jury and of the judge; of the political speaker and the voter; of the writer and teacher and their public; it forms part of the liberties for which every man with an idea of truth and every man with a pride in the dignity of his soul has fought since the beginning of society.

This struggle is today at its height; on its outcome will depend, among other great issues of a kindred nature, whether scientific detachment and the civilization pledged to respect and cultivate pure science shall perish from this earth.

91. From C. D. DARLINGTON's article, *Freedom and Responsibility in Academic Life* (1957)

[The great majority of scientists in the West are opposed to the planning by the State of research in pure science. It is policy in Great Britain and, perhaps to a lesser degree, in most other non-communist countries, that the universities should have the utmost possible freedom. That this too has its dangers is shown by the following.]

When the chair of physics at Oxford fell vacant in 1865 two candidates offered themselves: Hermann von Helmholtz, a notable German physicist, and Robert Clifton, an agreeable young mathematician with a gift for making instruments and a considerable estate in Lincolnshire. Of the electors two were divines who were not perhaps greatly interested in science and three were scientists who were certainly not interested in experiment. They elected Clifton. The new professor lived to a great age and for just fifty years he was successful in forbidding all new physical experiments in the Clarendon Laboratory. Helmholtz six years later became professor of physics in Berlin. He proved to be one of the great influences in the development of science, an influence stretching beyond his own country and beyond his own time.

The action of a small committee of electors who little knew what they were doing thus had a powerful influence on the development of the physical sciences both in Britain and in Germany and consequently on the course of the great wars of our own century.

SOURCES OF THE EXTRACTS

NUMERALS REFER TO EXTRACT NUMBERS

1. ARISTOTLE, *The Physics*, translated by P. H. Wicksteed and F. M. Cornford (Harvard University Press and Heinemann, 1934), Bk. I, ch. I, pp. 11–13.
2. PLUTARCH, *Lives*, Life of Marcellus, in the translation edited by John Dryden and revised by Arthur Hugh Clough, Everyman edition (Dent, London, and Dutton, New York), vol. I, pp. 471–4.
3. T. H. WHITE, *The Book of Beasts* (Jonathan Cape, 1954), pp. 7–9.
4. JONATHAN SWIFT, *Works* (London, 1784), vol. V, pp. 17 and 52.
5. COPERNICUS, 'De Revolutionibus Orbium Coelestium', translated by J. F. Dobson, assisted by S. Brodetsky, from Royal Astronomical Society, *Occasional Notes* (May 1947).
6. GALILEO, letter to Madame Christina of Lorraine, Grand Duchess of Tuscany, 'Concerning the Use of Biblical Quotations in Matters of Science', translated by Stillman Drake; from *Discoveries and Opinions of Galileo* (Doubleday and Company Inc., New York, 1957), pp. 175, 177, 181–2.
7. GALILEO, *Dialogue concerning the Two Chief World Systems— Ptolemaic and Copernican*, translated by Stillman Drake (University of California Press, 1935), pp. 321–6.
8. KEPLER, letters to Herwart; the first letter is from A. Koestler, *The Sleepwalkers* (Hutchinson, 1959), p. 340; the second letter is from Carola Baumgardt, *Johannes Kepler: Life and Letters* (Gollancz, 1952), p. 74.
9. ARISTOTLE, *The Physics*, as extract 1, Bk. VIII, ch. IV, pp. 315–17.
10. GALILEO, *Dialogue concerning the Two Chief World Systems— Ptolemaic and Copernican*, as extract 7, pp. 235–7.
11. GALILEO, *Dialogues concerning Two New Sciences*, translated by H. Crew and A. de Salvio (Evanston, Northwestern University Press), pp. 244–5.
12. NEWTON, *Principia*, in the translation from the Latin by Andrew Motte of 1729, revised by Florian Cajori (University of California Press, 1934), pp. 13–14, 550–2, 547.
13. ROBERT BOYLE, *A Disquisition about the Final Causes of Natural Things* (London, 1688), pp. 157–8.
14. WILLIAM HARVEY, *Exercitatio Anatomica de Motu Cordis et Sanguinis in Animalibus*, translated by Robert Willis (London, 1847), pp. 45–6.
15. TORRICELLI, letter to Michelangelo Ricci, from *The Physical Treatises of Pascal*, translated by I. H. B. and A. G. H. Spiers (Columbia University Press, New York, 1937), pp. 163–4.

16. MONSIEUR PERIER'S account to Monsieur Pascal, of the experiment performed on the Puy de Dôme, 19 September 1648, as extract 15, pp. 103–6.

17. OTTO VON GUERICKE, *Experimenta nova (ut vocantur) Magdeburgica de Vacuo Spatio*, translated by Martha Ornstein, from her book, *The Role of Scientific Societies in the Seventeenth Century* (University of Chicago Press, 1938), p. 51.

18. ROBERT BOYLE, *A Continuation of New Experiments touching the Spring and Weight of Air* (Oxford, 1669), pp. 42–5.

19. ROBERT HOOKE, *Micrographia* (London, 1665), Observation LIII.

20. LEEUWENHOEK'S letters, from Clifford Dobell, *Antony van Leeuwenhoek and his Little Animals* (John Bale, Sons and Danielsson Ltd, London, 1932), pp. 117–20, 168–70, 247–9.

21. FRANCIS BACON, *New Atlantis* (Cambridge, 1900), pp. 34–45 (abridged).

22. FRANCIS BACON, *Novum Organum*, translated by R. Ellis and James Spedding (Routledge, n.d.), p. 127.

23 RENÉ DESCARTES, 'Discours de la Méthode', translated by Elizabeth Anscombe and P. T. Geach in *Descartes, Philosophical Writings* (Nelson, 1954), pp. 15, 20–1.

24. THOMAS HOBBES, *Leviathan*, Everyman edition (Dent, London, and Dutton, New York), pp. 24, 82–3.

25. LAVOISIER, sealed note; from D. McKie, *Antoine Lavoisier* (Constable, 1952), pp. 74–5.

26. JOSEPH PRIESTLEY, 'Experiments and Observations on Different Kinds of Air', from Alembic Club Reprint No. 7, *The Discovery of Oxygen*, Part 1 (W. F. Clay, Edinburgh, 1894), pp. 6–7, 16–17, 54.

27. HENRY CAVENDISH, 'Experiments on Air', *Philosophical Transactions*, **74** (1784), pp. 134–5, 140, 150, 151–2.

28. LAVOISIER, *Traité Élémentaire de Chimie*, translated by R. Kerr (Edinburgh, 1790), pp. 78–81.

29. JAMES HUTTON, *Theory of the Earth* (Edinburgh, 1795); the first passage is from vol. 1, pp. 431 f., and the second passage from vol. 2, pp. 561–3.

30. WILLIAM SMITH, *Stratigraphical System of Organized Fossils* (London, 1817), pp. v, ix-x.

31. GEORGES CUVIER, *Discours sur la Théorie de la Terre*, translated by R. Jameson (Edinburgh, 1822), pp. 1–2, 6–7, 7–8, 14.

32. CHARLES LYELL, *Principles of Geology*, 10th ed. (John Murray, London, 1867), pp. 90–2.

33. JOHN LOCKE, *An Essay concerning Human Understanding*, Everyman edition (Dent, London, and Dutton, New York), pp. 26–7.

34. GEORGE BERKELEY, *The Principles of Human Knowledge*, 2nd ed., edited by T. E. Jessop (A. Brown and Sons Ltd, London, 1937), pp. 30–1.

35. DAVID HUME, 'An Enquiry concerning Human Understanding, from *David Hume, Theory of Knowledge*, ed. by D. C. Yalden-Thomson (Nelson, 1951), pp. 76–7, 159.

36. IMMANUEL KANT, *Critique of Pure Reason*, translated by N. K. Smith (Macmillan, 1953), pp. 127–8, 149–50.

37. DENIS DIDEROT, 'Conversation of a Philosopher with the Maréchale de X', from *Diderot, Selected Writings*, translated by Jean Stewart and Jonathan Kemp (Lawrence and Wishart, London, 1937), pp. 218, 228–9.

38. VOLTAIRE, 'A Treatise on Toleration', from *Selected Works of Voltaire*, translated by Joseph McCabe (Watts and Co., 1948), pp. 204–6.

39. *Conversations of Goethe with Eckermann and Soret*, translated by J. Oxenford (London, 1850), vol. I, pp. 106, 107, 108–10.

40. LUCRETIUS, *De Rerum Natura*, translated by Sir Robert Allison (Hatchards, 1925), pp. 46 and 53.

41. JOHN DALTON, *A New System of Chemical Philosophy* (London, 1808); the extract is taken from the edition of 1842, pp. 212–16.

42. AVOGADRO, 'Essay on a Manner of Determining the Relative Masses of the Elementary Molecules of Bodies, and the Proportions in which they enter into these Compounds', from Alembic Club Reprint No. 4, *Foundations of the Molecular Theory* (W. F. Clay, Edinburgh, 1893), pp. 28–9, 30, 33–4.

43. CHRISTIAAN HUYGENS, *Traité de la Lumière*, translated by S. P. Thompson (Macmillan, 1912), pp. 3–4.

44. NEWTON, *Opticks*, Dover Publications Inc. 1952 (based on the 4th ed. of 1730), pp. 362–3, 370, 347–8, 345–6.

45. THOMAS YOUNG, 'Reply to the animadversions of the Edinburgh Reviewers on some papers published in the Philosophical Transactions', from *Miscellaneous Works of the late Thomas Young*, ed. George Peacock (John Murray, 1855), vol. I, pp. 193–5, 201, 202–3.

46. FRESNEL, 'Mémoire sur la Diffraction de la Lumière', couronné par l'Académie des Sciences (1819), translated by A.E.E.M.; from *Œuvres complètes d'Augustin Fresnel*, tome premier (Paris, 1866), pp. 247–59.

47. JOULE, lecture 'On Matter, Living Force and Heat' (1847), from *The Scientific Papers of James Prescott Joule* (London, 1884), pp. 268–71.

48. J. R. MAYER, 'Remarks on the Forces of Inorganic Nature' (1842), translated from the German by G. C. Foster; from *Philosophical Transactions*, November 1862, pp. 375–7.

49. SADI CARNOT, *Réflexions sur la puissance motrice du feu* (1824), translated by R. H. Thurston (Macmillan, 1890), pp. 111–12, 113, 114–15.

50. FARADAY, 'On Lines of Force and the Field', taken from Michael

Faraday, *Experimental Researches in Electricity* (London, vol. III, 1855), pp. 438, 426, 447, 451.

51. J. C. MAXWELL, *A Treatise on Electricity and Magnetism* (Oxford, 1873), from the preface.

52. HEINRICH HERTZ, *Electric Waves*, published in German in 1892 and translated by D. E. Jones (Macmillan, 1893), pp. 20, 21, 27–8.

53. J. DUMAS and J. VON LIEBIG, 'Note sur l'état actuel de la Chimie organique', translated by A.E.E.M., from *Comptes Rendus de l'Académie des Sciences* (1837), tome v, pp. 567–70.

54. EDWARD FRANKLAND, 'On a New Series of Organic Bodies containing Metals', from *Philosophical Transactions*, **142** (1852), p. 440.

55. AUGUST KEKULÉ, *Über die Konstitution und die Metamorphosen der chemischen Verbindungen und über die chemische Natur des Kohlenstoffs*, translated by A.E.E.M., from Ostwald's *Klassiker*, no. 145, pp. 22–5.

56. AUGUST KEKULÉ, 'Untersuchungen über aromatische Verbindungen', translated by A.E.E.M., from Ostwald's *Klassiker*, no. 145, pp. 31–2.

57. J. B. LAMARCK, *Philosophie Zoologique*, translated by H. Elliott (Macmillan, 1914), pp. 107, 112–13, 117–18, 119–20, 122–3.

58. CHARLES DARWIN, *The Origin of Species* (John Murray, 6th ed. 1892), pp. 1, 14, 16, 28, 45, 46–7, 54, 60, 64, 97–8, 124, 134, 135, 136, 137, 276, 293, 294, 358–9, 360–1, 372–3, 387.

59. JOSEPH LISTER, 'On a New Method of Treating Compound Fracture, Abscess etc., with Observations on the Conditions of Suppuration', from *The Lancet*, 16 March 1867, p. 362.

60. LOUIS PASTEUR, 'Méthode pour prévenir la rage après morsure', translated by A.E.E.M., *Comptes Rendus de l'Académie des Sciences*, 26 October 1885, pp. 767–72.

61. ROBERT KOCH, *On Bacteriology and its Results*, translated by T. W. Hime (Baillière, Tindall and Cox, 1890), pp. 3–4, 11–12.

62. AUGUSTE COMTE, *Cours de Philosophie Positive*, translated by Harriet Martineau (London, 1853), pp. 1–2.

63. JOHN TYNDALL, 'Apology for the Belfast Address', from John Tyndall, *Fragments of Science* (Longmans Green and Co., 1899), vol. II, pp. 207–8, 218–19.

64. T. H. HUXLEY, 'On the Physical Basis of Life', from *Collected Essays*, vol. I (Macmillan, 1894), pp. 159–60.

65. ERNST HAECKEL, *Die Welträtsel*, translated by J. McCabe (Watts and Co., 1904), pp. 74, 82–3, 109.

66. HERBERT SPENCER, *First Principles* (Williams and Norgate, 6th ed. 1908), pp. 438–9.

67. AUGUST WEISMANN, *The Germ-Plasm, A Theory of Heredity*, translated from the German by W. N. Parker and H. Ronnfeldt (London, 1893), pp. xi, xii–xiii.

68. GREGOR MENDEL, paper on *Plant-Hybridisation*, translated by C. T. Druery and W. Bateson, *Mendel's Principles of Heredity* (Cambridge, 1909), pp. 309, 342–3, 344, 347.

69. T. H. MORGAN, *The Theory of the Gene* (Yale University Press, revised edition 1928), p. 25.

70. Statement by the Praesidium of the U.S.S.R. Academy of Sciences, quoted by Julian Huxley, *Nature, Lond.*, **163** (18 June 1949), pp. 935–6.

71. T. D. LYSENKO, 'The Situation in Biological Science', from *Agrobiology* (Foreign Languages Publishing House, Moscow, 1954), pp. 532–3, 537.

72. J. J. THOMSON, 'Cathode Rays', *Philosophical Magazine*, October 1897, pp. 293–4, 296–7, 310, 311, 312.

73. E. RUTHERFORD and F. SODDY, 'The Cause and Nature of Radioactivity', *Philosophical Magazine*, September 1902, pp. 394–5.

74. E. RUTHERFORD, 'Nuclear Constitution of Atoms', *Proceedings of the Royal Society*, Ser. A, **97** (1920), pp. 374–5, 379–80, 385.

75. ALBERT EINSTEIN, 'Lecture on the Theory of Relativity', from Albert Einstein, *Ideas and Opinions* (Alvin Redman Ltd, 1954), pp. 246–9.

76. MAX PLANCK, *The Origin and Development of the Quantum Theory*, translated by H. T. Clarke and L. Silberstein (Oxford, 1922), pp. 3–4, 11–13.

77. NIELS BOHR, 'Discussion with Einstein on Epistemological Problems in Atomic Physics', from *Albert Einstein, Philosopher-Scientist*, ed. P. A. Schilpp (Tudor Publishing Co., New York, fourth printing 1957), pp. 222–3.

78. WERNER HEISENBERG, 'Recent Changes in the Foundations of Exact Science', translated by F. C. Hayes, from Werner Heisenberg, *Philosophic Problems of Nuclear Science* (Faber and Faber, 1952), pp. 16–18.

79. EDWIN HUBBLE, *The Observational Approach to Cosmology* (Oxford, 1937), pp. 29–31.

80. ALBERT EINSTEIN, *The Meaning of Relativity* (Methuen, 4th ed. 1950), pp. 121–2.

81. H. BONDI, *Cosmology* (Cambridge, 1952), pp. 12, 140–1, 143–4.

82. ISAAC NEWTON, *Opticks*, as extract 44, p. 400.

83. ERNST MACH, 'The Economical Nature of Physical Enquiry', from Ernst Mach, *Popular Scientific Lectures* (Open Court Publishing Co., Chicago, 1895), pp. 192–3, 194, 198–9, 206–7.

84. HENRI POINCARÉ, *La Science et l'hypothèse*, translated by W.J.G. (Walter Scott Publishing Co. Ltd, 1905), pp. 160–1.

85. ALBERT EINSTEIN, *On the Method of Theoretical Physics*, as extract 75, pp. 271–4.

86. Sir ARTHUR EDDINGTON, *The Philosophy of Physical Science* (Cambridge, 1939), pp. 184–6.

87. RAPHAEL DEMOS, *Doubts about Empiricism*, from *Philosophy of Science* **14**, no. 3 (July 1947), pp. 203–4, 205–6, 207–8, 210, 215–16.
88. V. I. LENIN, *Materialism and Empirio-Criticism* (Lawrence and Wishart, London, 1948), pp. 95, 97, 99–100.
89. FREDERICK ENGELS, *Dialectics of Nature*, translated from the German by Clemens Dutt (Lawrence and Wishart, London, 1940), pp. 26–7.
90. M. POLANYI, letter on 'The Cultural Significance of Science', from *Nature, Lond.*, **147**, no. 3717, p. 119.
91. C. D. DARLINGTON, 'Freedom and Responsibility in Academic Life', *Bulletin of the Atomic Scientists*, **13**, no. 4 (April 1957), p. 133.